UNIFIED
OPTICAL SCANNING
TECHNOLOGY

UNIFIED
OPTICAL SCANNING
TECHNOLOGY

LEO BEISER

IEEE PRESS

JOHN WILEY & SONS PUBLICATION

Published by John Wiley & Sons, Inc., Hoboken, New Jersey.
Published simultaneously in Canada.

For general information on our other products and services please contact our Customer Care Department within the U.S. at 877-762-2974, outside the U.S. at 317-572-3993 or fax 317-572-4002.

Wiley also publishes its books in a variety of electronic formats. Some content that appears in print, however, may not be available in electronic format.

Library of Congress Cataloging-in-Publication Data:

Beiser, Leo.
    Unified optical scanning technology / Leo Beiser.
       p.   cm.
    Includes bibliographical references and index.
    "A Wiley-Interscience publication."
    ISBN 0-471-31654-7 (cloth)
    1. Optical scanners.   2. Optical storage devices.   I. Title.

TK7882.S3 .B49 2003
006.6'2—dc21   2002032381

10 9 8 7 6 5 4 3 2 1

To Edith
My Golden Coherent Light

CONTENTS

Preface   **xi**

**1 INTRODUCTION—TECHNOLOGY OVERVIEW AND
    UNIFYING PRINCIPLES**   **1**

   1.1 Optical Scanning Characteristics and Disciplines / 1

   1.2 Active and Passive Scanning / 3

      1.2.1 Conjugate Image Representations / 4

      1.2.2 Retroreflection and Double-Pass Systems / 5

   1.3 Input, Output, and Remote Sensing Systems / 8

   1.4 Optical and Resolution Invariants; Optical Transfer / 9

   1.5 System Architecture / 12

      1.5.1 Objective Lens Relationships / 13

2 SCANNING THEORY AND PROCESSES   **19**

   2.1 The Point Spread Function and Its Convolution / 19

      2.1.1 PSF Developed from Uniform Illumination of an
                    Aperture / 20

      2.1.2 PSF Developed from Aperture Illumination with a
                    Gaussian Distribution / 22

      2.1.3 Scanning—Controlled Movement of the PSF; Its
                    Convolution / 25

2.2   Quantized or Digitized Scan / 27

2.2.1   The Sampling Criterion / 28

2.3   Gaussian Beam Propagation / 31

2.3.1   Representation and Development of the Gaussian Beam / 31

2.3.2   Gaussian Beam Focusing Characteristics / 35

2.4   Scanned Quality Criteria and the Modulation Transfer Function / 37

2.4.1   The Fourier Transform / 38

2.4.2   The Modulation Transfer Function / 40

3   SCANNED RESOLUTION   **45**

3.1   Influence and Significance of Scanned Resolution / 45

3.1.1   Basis of Scanned Resolution / 45

3.1.2   Resolution Nomograph / 48

3.2   Aperture Shape Factor / 50

3.2.1   Uniformly Illuminated Apertures / 50

3.2.2   Summary of Apertue Shape Factors / 52

3.3   The Resolution Equation, the Resolution Invariant, and Beam Propagation / 54

3.3.1   Propagation of Noise and Error Components / 54

3.4   Augmented Resolution / 56

3.4.1   Radial Symmetry and Scan Magnification / 57

3.4.2   Augmented Resolution for Holographic Scanners / 60

3.5   Resolution in Passive and Remote Sensing Systems / 61

4   SCANNER DEVICES AND TECHNIQUES   **63**

4.1   Scanner Technology Organization / 63

4.2   High-Inertia Scanning / 65

4.3   Rotating Polygons / 65

4.3.1   Distinctions Between Pyramidal and Prismatic Polygons / 66

4.3.2   Duty Cycle / 67

4.3.3   Over- and Underillumination (Over- and Underfilling) of the Facet / 68

4.3.4   Facet Tracking / 69

4.3.5   Design Considerations / 69

4.3.6   Passive Scanning for Remote Sensing / 83

4.4   Holographic Scanners / 85

4.4.1   Scanner Configurations and Characteristics / 87

4.4.2   Implementation of Holographic Scanners / 92

4.5   Oscillatory (Vibrational) Scanners / 100

4.5.1   The Galvanometric Scanner / 101

4.5.2   The Resonant Scanner / 103

4.5.3   Suspension Systems and Position Control / 104

4.5.4   The Fast-Steering Mirror / 105

4.5.5   The Fiber Optic Scanner / 106

4.6   Scanner-Lens Relationships / 108

4.6.1   Scanner-Lens Architecture / 108

4.6.2   Double-Pass Architecture / 109

4.6.3   Aperture Relaying / 111

4.6.4   Lens Relationships for Control of Deflection Error / 112

4.7   Low-Inertia Scanning / 112

4.8   Acoustooptic Scanners / 113

4.8.1   Operating Principles / 113

4.8.2   Fundamental Characteristics / 116

4.8.3   Alternate Acoustooptic Deflection Techniques / 117

4.9   Electrooptic (Gradient) Scanners / 124

4.9.1   Implementation Methods / 126

4.9.2   Drive Power / 128

4.10 Agile Beam Steering / 128

4.10.1 Phased Array Technology / 129

4.10.2 Decentered Microlens Arrays / 139

4.10.3 Summary of Agile Beam Steering / 144

5 CONTROL OF SCANNER BEAM MISPLACEMENT   147

5.1   Cross-Scan Error and Its Correction / 148

5.1.1   General Considerations and Available Methods / 148

5.1.2   Passive Methods / 150

5.2   The Ghost Image and Its Elimination / 155

    5.2.1   Skew Beam Method of Ghost Elimination / 156

    5.2.2   Beam Offset Method of Ghost Elimination / 156

6   SUMMARY—MAJOR SCANNER CHARACTERISTICS   **161**

6.1   Comparison of Major Scanner Types / 164

References   **169**

Index   **179**

PREFACE

What is *Optical scanning?* Let me offer a definition. *Optical scanning is a systematic articulation of light to provide information transfer.* Then, what is *information?* Consider that it means simply *organized knowledge.* What were the first optical scanning innovations? *Semaphores?* Or, how about, *smoke signals?* That is, *amplitude modulation of scattered light from controlled groups of particulate aerosol, which disperse as they rise, as observed from a distance.* This is a form of serial information transfer.

The two principal forms of optical information transfer are identified as *serial* and *parallel.* The parallel form is represented classically by fields such as *telescopy, microscopy,* and *photography*—effectively rendering simultaneous transport of the information. The serial form is dominated currently by familiar fields such as television, E-mail, and the reproduction of the graphic arts and sounds, utilizing an arranged timed series of pixels, pels, or voxels. This temporal factor is of basic interest, for it allows (via the optical/electrical transducer—modulator and detector), efficient electronic processing. It is because of this fundamental flexibility that optical scanning has engendered such a commanding role for information transfer. As is surely recognized, many of the classic parallel systems have evolved to either hybrid serial/parallel or fully serial utilizing various forms of spatial scanning to gain the operational advantages in data manipulation and transfer. Adept development requires, however, dedicated attention to the breadth of the disciplines that may be integrated into these typically diverse systems. An objective of this work is to render a unified orientation to these fundamental and often secluded related factors.

Optical scanning may be conducted with many "shades" of "light," ranging spectrally from UV through the visible to IR and exhibiting incoherent to coherent order. Although the transfer of optical information that is outside that range of electromagnetic radiation is perfectly valid, as notably by CT (X-ray) scans, they are limited in the angular scan flexibility manifest in the dominant forms of optical scanning. Similarly, with the introduction of the laser in the early 1960s, exhibiting its extremely disciplined propagation characteristics, active optical scanning is typically of the laser beam, identified compactly as *laser scanning*.

One important exception is the field of *remote sensing*. Here, passive *incoherent* radiation from a remote source is collected optically *at a distance*. Its spatial and spectral composition is analyzed by scanning/detection devices to form organized patterns of pixels for subsequent reconstitution. Significant for us is that the scanning devices for each field—operating from very independent viewpoints—exhibit remarkably common factors. This has been highlighted only recently in a publication by this author and R. Barry Johnson, with the objective of unification of the technology. The background and comparative disciplines are expanded in this work.

Other aspects of commonality are the fundamental analogies between different disciplines, as in optical and electronic scanning theory, and in the sampling requirements for effective image, data, and acoustic reconstruction. Recurrent cross-referencing of related factors appears in different portions of this volume. Included is attention to significant arcane techniques such as the Scophony system and several other enlightening acoustooptic scanning methods. Another realization of this work is the periodic heuristic clarification of the fundamentals, with little dependence on the use of detailed mathematical affirmation—which may be accessed readily in the referenced literature.

In recognizing a scanning technology with an extensive background, which has been enhanced under government support for over a decade, this volume offers the first general publication on the advancing field known as *agile beam steering*. Germinating over the past three decades (reported by this author in 1974), it has attained serious R&D attention for broad application. Agile beam steering is reviewed and presented comprehensively as another unifying dimension.

Rendering now only the thematic highlights, it is noteworthy that the final chapter in this volume summarizes the principal thrust of each chapter. It closes with a unique and potentially controversial independent compilation of the major scanner types, avoiding the almost inevitable preferences that can appear in summaries published by well-meaning researchers, developers, or manufacturers.

Taking a moment to reflect on and recognize the people who have nurtured this work, I extend my unlimited admiration and thanks first to my loving wife, Edith, who has tolerated the diversions, machinations, interruptions, and seclusions, for the sake of whatever accomplishment this may have represented in our minds. Since I have retired officially recently, after 25 years of independent consultation practice, with many prior rewarding writing experiences behind me, and now looking forward to long-delayed relaxation, I'd be *out of my mind* to *even think* of doing this again.

Valued support and guidance was rendered in astute review of several chapters by R. Barry Johnson of Optical ETC. Inc., with whom I have had the pleasure of working closely in prior co-authorship and in many exhilarating discussions. Another esteemed review was conductd by Ed Watson of the Air Force Research Labs, rendering revealing and supportive comment on my coverage of a technology of his renowned specialty. The collective total of my professional relationships has surely nurtured this work. My colleague Gerald Marshall, who, after joining the consultation field, also joined me in cochairing many of our SPIE Technical Conferences and rendering valued diversity. More recently, this resource was enhanced with the cochairmanship of Steven Sagan, who adds the inspiration of youth and the breadth of lens design to our conference skills. Many more—far too many to identify—have surely formed a composite inspiration. You know who you are. Special regard is accorded to George Telecki, Associate Publisher at John Wiley & Sons, Inc., who has provided much appreciated patience and understanding in his guidance of this work.

My sincere thanks to all.

Leo Beiser
Flushing, New York

CHAPTER 1

INTRODUCTION—TECHNOLOGY OVERVIEW AND UNIFYING PRINCIPLES

1.1 OPTICAL SCANNING CHARACTERISTICS AND DISCIPLINES

Optical scanning serves as an information converter. It transforms a spatial function (such as an image) into a time domain (a signal) to adapt it to electronic processing. Or, in reciprocal form, scanning decodes a series of signal impulses to assemble its image or spatial function. Familiar examples are the systems of television or facsimile. Optical scanning may also be confined to the (nonimaging) data domain, analog or digital, as in the field of data storage and retrieval. Here, it interprets the elemental optical changes in a storage medium as an electrical signal. Or it renders the reverse process of transforming an electrical data stream into detectable changes in a storage medium. Familiar examples are bar code or compact disk (CD) recording and reading. In these ways, optical scanning serves as an information encoder or decoder, truly a key to advanced information transfer.

We can express this conversion process in terms of transformations between spatial orientations (which may be a function of time) and the time. In "moving images," for example, the spatial orientations s are themselves a function of the time. Letting $s_{c,t}$ represent the three spatial coordinates ($c = 1, 2, 3$) and t the time, one may write,

$$f(s_{c,t}) \longleftrightarrow g(t) \qquad (1\text{-}1)$$

to symbolize the reciprocal transformations by means of optical scanning of the spatial function f to or from the time function g.

Portions of these operations are performed by familiar static components. For example, in a bar code reader, in addition to the optical scanner that directs illumination to the bar code elements in rapid succession, there appear intervening optics that shape the scanned output beam, optics that collect some of the scattered flux from the bar code and concentrate it upon a detector that transduces this varying flux into a corresponding elemental signal for subsequent processing. It is the combination of such fundamental components (e.g., light source, scanner, optics, modulator, and detector) that forms a subsystem called a digitizer, printer, or recorder. The arrangement of these components is discussed in Sections 1.3 and 1.5.

Optical scanning applies as well to the important complementary field of image detection, which is sometimes performed unobtrusively. In remote sensing, the flux radiating from a remote object is sampled and directed to a detector for conversion to a corresponding electrical signal. Because the scanning and optical components deployed in both fields (active and passive) are often very similar fundamentally, significant attention is devoted here to their unification. The reciprocal properties of these systems are introduced in Section 1.2 and are then exemplified. Other unifying principles, such as the resolution invariant, are represented in this work.

In an information handling system, optical scanning appears most often in the peripheral equipment, in the output or input devices that either enter data or interpret data for electronic manipulation. In this manner, optical scanning is analogous to the microphone and earphone of an audio system, operating in the optical domain on spatial information, rather than on acoustic waves. One can see how the characteristics and disciplines of optical scanning escalate in complexity to affirm that "a picture is worth a thousand words."

A pattern of unifying principles, pedagogic analogies, and rarely posed topics represented in this work may be listed usefully, in approximate sequence of introduction, to include—

- The reciprocal space-time concept
- Complementary active-passive scanning
- Conjugate processes in remote sensing
- The resolution invariant and its significance
- Electrical-Optical signal-image theory
- Analogous low-pass filter and scanned resolution loss

- Analogous oversampling criteria in different fields
- Analogous Fourier transform and spectrum analysis
- Analogous conventional and holographic scanners
- Analogous Bragg diffraction and prism refraction
- Analogous radar and optical phased arrays
- Rarely posed image rotation and derotation
- The fast steering mirror—newly developed
- The resonant scanner classified as "high inertia"
- Scanner-lens architecture options
- Rarely posed, aperture relaying
- The Scophony process—rarely posed
- Rarely posed traveling lens and chirp deflectors
- Propagation of noise and error components
- Agile beam steering—new and comprehensive
- Cross-scan error control—its own chapter
- Ghost image elimination, newly rendered

The main thrust of this volume is to facilitate judgment for selecting, from a broad range of optical scanning techniques and architectures, the most effective methods for design and further development of information transfer systems. Insight for creative advancement is fostered with the unification of some diverse and arcane concepts. This work ends with Chapter 6, which summarizes the content and charts the characteristics of the principal operational scanner prototypes, judgment seldom rendered with independent consideration.

1.2 ACTIVE AND PASSIVE SCANNING

This section expands on the two complementary forms of optical scanning: **active** and **passive**. Active scanning directs illuminating flux *to different portions* of an object, and passive scanning samples and directs radiant flux *from different portions* of a remote object to a *detector*. Although the scanning processes are often similar and typically very fast, they propagate radiant flux along reciprocal paths. Active scanning has been described earlier as "flying spot scanning," often implemented now as laser scanning. It moves a point of light in a regular pattern across an information-bearing medium or recording surface. The complementary system of passive scanning is exemplified by remote sensing of an infrared image. It serves a function similar to that of focal plane array image sensing, requiring, however, focusing on a single (or a few) optimized radiation detector(s) rather than on an array of many (to millions) addressable detectors. An objective of this work is to unify further the two major fields of active and passive scanning [B&J]. Developed by essentially independent research communities for unobtrusive image acquisition, they were sheltered because of their secure applications. Active laser scanning was initiated to advance reconnaissance phototransmission [Bei1, Appendix 1], whereas passive scanning served to enhance image sensing, notably in the IR spectrum [Wol].

1.2.1 Conjugate Image Representations

The concepts of active and passive scanning are unified, utilizing the principles of conjugate point imaging and ray reversibility [Lev]. Introduced later are two related forms of active scanning, namely, retroreflection and double-pass operation.

Consider Figure 1.1; a classic representation of conjugate imaging, with the added provision for optical beam scanning [Bei2]. Rays from a reference (object) point at P_o are transferred lenticularly to its conjugate image point at P_i, while encountering regions that provide means for scanning the ray group. With laser illumination forming the object point P_o, Figure 1.1 represents active scanning. Its rays propagate from left to right through the objective lens to focus at the image points P_i. If the input beam is collimated, the object point P_o appears effectively at $-\infty$.

Establishing the objective lens at a reference position in the propagation path, scanning of the active beam *before* the lens is identified as *preobjective scanning* and scanning after the lens as *postobjective*

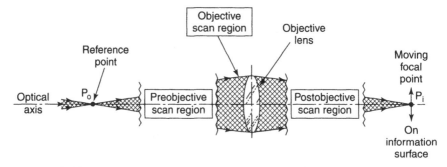

Fig. 1.1 Conjugate imaging system representing active (flying spot) scanning. Ray directions (left to right) from object point P_o to scanned image points P_i. Active scan regions are identified.

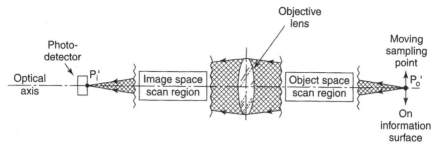

Fig. 1.2 Conjugate imaging system representing passive scanning. Ray directions (right to left) from radiating object points P_o' to image point P_i' for signal extraction. Passive scan regions are identified.

scanning [Bei2]. The loci traced by the scanned image points are different, and they are discussed subsequently. Translation of the objective lens *or* the information surface (or both) in a direction transverse to the axis is identified as *objective scanning* [Bei2]. These definitions and their architectural consequences are illustrated and expanded in Section 1.5.

Consider now Figure 1.2, a schematic representation of passive scanning. With the same property of conjugate point imaging, and with regions that are analogus to those in Figure 1.1, the main distinction from active scanning is that the ray directions are reversed. They now propagate from right to left, from the new object point at P_o' to the new image point at P_i'. The preobjective and postobjective nomenclature of active scanning have their counterparts here, defined as *image space* and *object space* scanning, respectively [B&J]. Operationally, the object space could extend over a very great distance, allowing for the imaging of substantially collimated flux entering the system.

1.2.2 Retroreflection and Double-Pass Systems

The general representations of active and passive scanning are here advanced to two important variations to active scanning: retroreflection (or retrocollection) and double-pass operation. Although distinct in utility, they exhibit architectural similarities represented here and in Section 1.5.

1.2.2.1 Retroreflection In retroreflection or retrocollection, an information medium is scanned actively to detect and interpret its uniquely scattered flux. More specialized than flying spot scanning, retroreflection concentrates on "reading" primarily the reversed-path

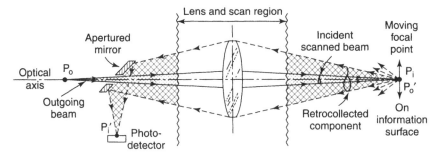

Fig. 1.3 Illustration of retroreflection scanning, incorporating Figs. 1.1 and 1.2. Outgoing beam (solid lines) propagates left to right from object P_o to be deflected and imaged at P_i. This becomes effectively a new source P_o' radiating right to left (dashed lines), a portion of which is retrocollected at the lens-scanner aperture, to be descanned and directed to reimage as P_i' on the fixed photodetector.

(retroreflected) scatter or specular component, rather than detecting scatter in other directions. Retroreflection provides an advantageous and simple architecture that also "descans" the return beam, directing it automatically and efficiently to a small photodetector. This reversed-path flux represents more of a specularly reflected component when the optical system is telecentric, in which all scanned beams are incident substantially normal to the flat image surface.

Figure 1.3 is a schematic representation of a retroreflection system. It consists essentially of both Figures 1.1 and 1.2 above, combining the reciprocal paths of active and passive scanning. The *active* component is represented by the narrower-beam outgoing group of rays (solid lines). It emanates from object pont P_o, propagating to the "right" through a small central aperture in mirror M, through the lens and scanners to illuminate image point P_i. This point is now considered an object point P_o' of an effective radiating source. The retrocollected portion of this scattered radiation from P_o' returns (to the "left") as the larger-aperture ray group (dashed lines), toward the lens and scanner or beam deflector. The lens then collects and the deflector descans and returns the beam to the axis of origin. The larger-diameter portion of the return beam is intercepted by the mirror, which reflects and redirects it to focus at a new image point P_i' on a fixed photodetector, achieving efficient signal extraction. Alternates to the apertured mirror include the use of a partially silvered mirror (beam splitter) or polarization separation or inverting the role of the apertured mirror, the use of a small "turning" mirror (at the aperture location) that is angled to intercept and redirect to on-axis an initially off-axis small-diameter input beam.

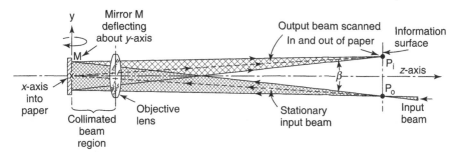

Fig. 1.4 Illustration of double-pass operation. The objective lens of Fig. 1.1 is effectively, split in half and separated such that the folding and deflecting mirror M is positioned in the center of the now collimated region. The lens now serves in double pass, to first transfer fixed source point at P_o to illuminate mirror M, which, in preobjective scanning, forms the locus of image points P_i in and out of the paper.

Then, almost all of the larger return beam (surrounding the small mirror) is brought to focus on to a fixed photodetector that is now on-axis. (Further examples of retroreflection or retrocollection appear in Section 4.4.2 related to Fig. 4.15, and in Section 5.2.2 regarding Fig. 5.7.)

1.2.2.2 Double-Pass Operation

In a double-pass system [Bei2], the basic Figure 1.1 is modified by dividing the objective lens effectively in half and folding the system about its nodal center with a mirror, to use the lens twice. As illustrated in Figure 1.4, this places the conjugate image points P_i and P_o on the same side. The single objective lens (with reflection from mirror M) now serves in double-pass, to first collimate the input beam from P_o on to mirror M and then, acting as a flat-field lens, to focus the output beam to P_i. The scanning is performed by mirror M, which both folds and deflects the beam in a preobjective scan mode. An alternative to this input/output separation technique is shown in Figure 4.21 and is discussed in Section 4.6.2. Although appearing similar to Figure 1.3, double-pass operation is distinguished from retroreflection by the insertion of mirror reflection in the pupil region of the objective lens rather than in the image region.

In this form, M is a mirrored deflector that is actuated angularly about the vertical (y) axis to deflect the image point P_i along a path that is essentially perpendicular to the paper. The angular separation β between input and output beams (in the y–z plane) is nominally small, to impose minimal bow (sag in the y-direction) to the locus of image points P_i. This is detailed further in Section 4.6.2. If P_o and P_i (when undeflected) were on the z-axis, the bow would be nulled and the output beam would be confined to the x–z plane, normal to the plane of the paper.

A virtue of double-pass operation is its space and component conservation, whereby the objective lens serves as the collimator lens of a beam expander for the fixed input beam and also as the flat-fielding lens that linearizes the scanned locus of output beam points P_i. When operating with near-normal landing of the collimated beam on (undeflected) mirror M, this also minimizes the size of this deflecting component—by minimizing the angle of off-axis arrival of the input beam (as would otherwise be required to bypass the flat-field lens). Additionally, this minimizes the pupil relief distance (between the scanner and the first surface) of the flat-field lens, minimizing lens sizes for a given scan angle. Further discussion and illustration of double-pass operation appear in Sections 4.3.5.3 and 4.6.2. Noteworthy here are the commonality and flexibility of the conjugate image representation of optical scanning.

1.3 INPUT, OUTPUT, AND REMOTE SENSING SYSTEMS

The conjugate optical systems introduced earlier reenforce the functional unity of input and output systems, in which similar scanners and optics are utilized. Table 1.1 identifies some prominent applications. The first two columns illustrate forms of (active) flying spot scanners, represented in Figures 1.1 and 1.4. Those in the Outputs column of Table 1.1 require only the upper branch of Figure 1.5, providing scanned and modulated spot illumination on the information medium, as in typical

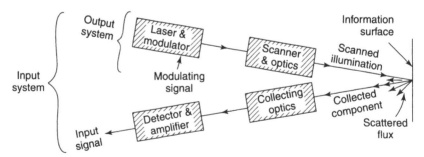

Fig. 1.5 Block diagram of generic input and output systems showing common major components. Upper branch represents output system, and a combination of both branches represents active Input systems. The lower branch alone, with added scanner, would represent passive scanning.

TABLE 1.1   Some Applications of Optical Scanning Technology

Active (Laser) Scanning		Passive (remote sensing)
Outputs (from data system)	Inputs (to data system)	
Image recording/printing	Image digitizing	Night vision
Color image reproduction	Bar code reading	Surveillance
Medical image recording	Optical inspection	Gun sighting
Data marking and	Confocal microscopy	Weather monitoring
engraving	Optical character	Missile tracking
Microimage recording	recognition	Earth resources sensing
Optical data storage	Graphic arts camera	Medical image sensing
Phototypesetting	Color separation	Thermal fault detection
Earth resources recording	Robot vision	Nondestructive testing
Data/image display	Laser radar	Forest fire monitoring

recording applications. The modulating device may be integral with or separate from the laser. The scanners in the Inputs column require both the scanned spot illumination from the upper branch (without signal modulation) and the signal extraction lower branch, collecting and detecting the scattered flux.

The last column in Table 1.1 exemplifies passive scanning, dominated by the field of remote sensing, primarily in the infrared spectrum. Although these listed applications generally sense and/or image self-radiating objects, this passive category includes the option, if expedient, of benefiting from external illumination upon the object. From the discussions of Figures 1.1 and 1.2, it will be apparent that the lower branch of Figure 1.5 alone (with added deflector) represents passive scanning. A distinction between the collecting optics of the input system in Figure 1.5 and the imaging optics of remote sensing is that the collecting optics of Figure 1.5 is usually located advantageously close to the object being sensed. This option is seldom available in remote sensing, where optimizing detection sensitivity over an extended object space is one of the significant design challenges.

1.4 OPTICAL AND RESOLUTION INVARIANTS; OPTICAL TRANSFER

A most revealing relationship in optical scanning systems is an adaptation of the optical invariant to the resolution invariant [Bei3]. The optical invariant is known by several names, notably as the Lagrange invariant [Good, Lev]. As represented in Figure 1.6, it relates the object

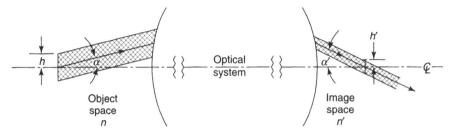

Fig. 1.6 Illustration of the Lagrange invariant, $L = nh\alpha = n'h'\alpha'$, in which the unprimed factors appear in object space and the primed factors appear in image space, coupled through a ray-continuous coaxial optical system. Notably, when $n = n'$, then $h\alpha = h'\alpha'$, having consequences for scanned resolution and the propagation of angular errors.

height h and the paraxial (where $\sin\alpha \to \alpha$) input ray angle α to the image height h' and the paraxial output ray angle α'. Allowing also for the refractive index n in object space and the refractive index n' in image space, the Lagrange invariant is expressed as

$$L = nh\alpha = n'h'\alpha' \qquad (1\text{-}2)$$

Thus, in a space of common refractive index, the product of the object height and its input propagation angle is equal to the product of the image height and its propagation angle. This first-order paraxial approximation includes tandem optical elements or stages (so long as they are ray-contiguous coaxial spherical surfaces of revolution). Although not noted explicitly elsewhere, it also applies to the special case of the meridional plane of anamorphic (e.g., cylindrical) optics, an important feature in some optical scanning systems (see Chapter 5).

The Lagrange invariant merits special attention here because of its revealing adaptation to scanned resolution—a topic to which Chapter 3 is devoted. To demonstrate this relationship, we introduce first some characteristics of scanned resolution. In contrast to the popular representation of "hard copy" resolution in "dots per inch," or in similar "number/unit distance" ratios, (e.g., photographic lines/mm), scanning (which may, of course, create the hard copy) conveys to its **full format** a total number of resolvable elements along its scanned line. Thus scanned resolution ascribes to the line a **total number of N elements**. And, because scanning always involves the time, the number of resolution elements per unit time relates directly to signal bandwidth. Television resolution, for example, is expressed in the number of lines per full frame. In remote sensing, the line-scanned resolution is expressed (Chapter 3) as the number of elements $N = \Theta/\Delta\Theta$, the full-field angle

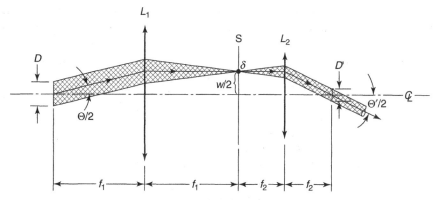

Fig. 1.7 An afocal optical system illustrating Fig. 1.6 and the resolution invariant $I_N = \Theta D = \Theta'D'$. Telescopic transfer (as a beam compressor or an optical relay) of the scanned small angle $\Theta/2$ from aperture D to larger output angle $\Theta'/2$ at smaller output aperture D', such that $\Theta'/\Theta = D/D' = f_1/f_2$.

Θ divided by the smallest detectable angle $\Delta\Theta$ (instantaneous field of view).

When $\Delta\Theta$ is the diffraction-limited beam divergence from a source diameter D, Chapter 3 reveals simply that the resolution N conveyed by an angularly scanned beam is proportional to the product of two main variables: the total angle Θ through which the beam scans and the width D of the beam at the scanner. Thus, letting k denote a system constant, the resolution is represented by

$$N = k(\Theta D) \qquad (1\text{-}3)$$

Consider the first half of Figure 1.7 (to surface S). A scanning source directs a beam of diameter D through half-angle $\Theta/2$ to lens L_1 (having a focal length f_1). The lens focuses the beam to spot size δ on surface S, covering $\frac{1}{2}$ of the format width $W/2$ with adjacent spots. The **number** N of such spots over the format width W is established **at the deflector**; functionally according to Equation 1-3. Operational details are provided in Chapter 3. Figure 1.7 is also involved in subsequent chapters, relating to beam expansion/compression and optical relaying.

To express the resolution invariant I_N, let the αs of Equation 1-2 be Θ and Θ' respectively, and the h values of Equation 1-2 be D and D', respectively. Then, with $n = n'$ and with k constant, Equation 1-3 forms the resolution invariant [Bei2]

$$I_N = \Theta D = \Theta'D' \qquad (1\text{-}4)$$

The resolution (N elements per scan) is conserved throughout optical transfer. That is, **scanned resolution N is independent of the optics following the deflector**. The resolution invariant of Equation 1-4 is manifest in the second half of Figure 1.7, which shows lens L_2 reforming the beam to a smaller diameter D' at the output and executing a proportionately larger scan angle $\Theta'/2$, such that their product remains equal to that at the input. One may then focus the scanned output beam and develop the same resolution N over the output format. Notably, the "flat-field" lens following the deflector in active systems (which determines the format, spot size, linearity, and other important characteristics) **does not determine resolution N**. Except for its possible aberration (of wave front integrity or spot quality), scanned resolution N is determined **at the deflector** and not by the transferring optics.* The format and spot size established by the lens are not independent, for their values form a fixed ratio of N = scan length/spot size; as determined from the basic relation $N = \theta/\Delta\theta$. Chapter 3 is devoted to the important topic of scanned resolution. Definition of the spot size as a function of beam shape and its measurement points, consistant with the explicit forms of Equation 1-3, is provided in Chapters 2 and 3.

1.5   SYSTEM ARCHITECTURE

Some architectural options were introduced in the sections above. The discussion is discussed here regarding the spatial and optical relation-

---

* A related optical invariant is the *étendue* [Hob] or *optical throughput*, given by

$$\acute{E} = n^2A\Omega = n'^2A'\Omega' \qquad (1\text{-}5)$$

It is a two-dimensional function of the source *area A* and its radiating *solid angle Ω* representing input-to-output *power transfer*. The *étendue* is most descriptive of incoherent radiation from an object that reaches an image. Coherent propagation is so confined and analytic (see Chapters 2 and 3) that the control of coherent power throughput entails, typically, the minimizing of any truncation of the beam along its propagation path.

Considering Equation 1-5 for a uniform index medium, and assuming an isotropic source, $A \propto D^2$ and $\Omega \propto \theta^2$, where θ represents the one-dimensional radiating angle, then one may write

$$\acute{e} \equiv \sqrt{\acute{E}} \propto \theta D = \theta'D' \qquad (1\text{-}6)$$

Rather than forming an incongruous "one-dimensional optical throughput," it leads to the resolution invariant of Equation 1-4, representing consistent information transfer.

ships among the system components and the consequences of these arrangements.

1.5.1 Objective Lens Relationships

Section 1.2.1 described the active systems represented by Figure 1.1 utilizing three unique orientations of the scanning component with respect to the objective lens. They were termed, preobjective, postobjective, and Objective. Each type renders distinct characteristics, primarily with regard to the loci of the scanned spot and the complexity of the associated optics. The characteristics of passive systems represented by Figure 1.2 are compared with those of Figure 1.1.

1.5.1.1 Preobjective Scanning Figure 1.8 illustrates a typical component arrangement of preobjective scanning. Mirror M is oscillated or rotated through angle Φ about axis A, which is parallel to the y-axis. On illumination of the mirror with a beam of diameter D, the reflected principal rays execute the total scan angle Θ in the x–z plane and are focused by the objective lens to form the indicated scan line. This line lies in the x–z plane (unbowed in the y-direction) when the mirror normal and the input beam are both in the x–z plane. The straightness of the line ("flatness" in the x–y plane) and its linearity (increments of spot displacement along x being proportional to increments of scan angle Θ) are two of the important parameters that are controlled by the lens. Considerable challenge develops for design of

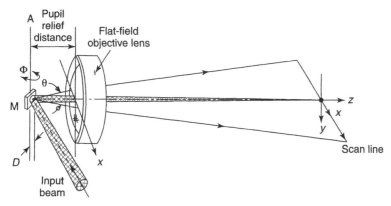

Fig. 1.8   Representation of preobjective scanning by mirror M through angle Φ about axis A. Input beam is scanned through angle Θ and is directed through the objective lens to focus the beam along the scan line. Only principal rays of the deflected beam are illustrated, for simplicity. [Bei 8].

this "flat-field" lens for $\Theta > \pm 30°$, while providing appropriate image flatness and linearity; especially for $N > 20{,}000$.

The pupil relief distance noted in Figure 1.8 and introduced at the end of Section 1.2.2.2, is another significant design parameter. It merits comparison with alternate configurations. This distance is determined by the diameters of the flat-field lens housing and the input beam and the angle of this beam with respect to the z-axis. As shown in Figure 1.8, adequate clearance must be provided between the input beam and the edge of the lens housing to avoid beam truncation (vignetting). Design procedures appear in Chapter 4. Increasing the pupil relief distance while maintaining scan angle Θ increases the resulting size (and hence cost) of the flat-field lens. To reduce this distance requires increasing the angle of the input beam with respect to the z-axis, which forces an increased size of the deflecting device (mirror M) to subtend the beam. This is especially significant when the mirror M is one of many on a prismatic polygon. It is shown (Chapter 4) that the polygon diameter is inversely proportional to the cosine of a related angle α, whose cosine reduces rapidly for incident beam angles greater than 45°. This in turn demands a rapid enlargement of the polygon diameter.

An alternate to this configuration, which remains preobjective, is to arrange the components as illustrated in Figure 1.9. The mirror M is now disposed nominally at 45° to its shaft rotation axis A, and the input beam is now coaxial with the shaft. Although operational differences exist in comparison to the former arrangement (detailed in Chapter 4), the above interference between the input beam and the flat-field lens disappears. Essentially unlimited scan angle is available (within the limitations of a flat-field lens), along with minimal pupil relief distance and

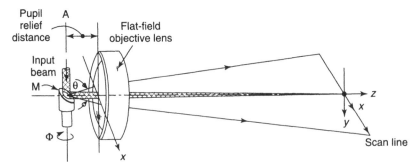

Fig. 1.9 Alternate scanner orientation for preobjective scanning with input beam parallel to axis A. Shaft suppports monogon mirror M at nominal 45° angle to rotational axis, allowing shortened pupil relief distance. Also, operating in radial symmetry, $d\Theta/d\Phi = 1$, whereas in Fig. 1.8, $d\Theta/d\Phi = 2$.

no demand for an increased mirror size. As the mirror M in Figure 1.8 is extendable to a prismatic polygon, so the single mirror in Figure 1.9 on a "monogon" scanner is extendable to a pyramidal polygon (Chapter 4). It is also extendable to a family of scanners that provide correction for line placement errors (Chapter 5). Also similar to the disappearance of the above interference between the input beam and the flat-field lens is a property of some holographic scanners that operate as rotating transmission gratings (Chapter 4). Another distinction between this class of pyramidal scanners and prismatic scanners relates to their typical orientation with respect to the incoming flux, described as "radial symmetry." This fundamental characteristic is discussed in Section 3.4.1. A consequence is that scanners such as monogons or pyramidal polygons that operate in radial symmetry also rotate the image about its projected axis. Similarly, in passive scanning, they rotate the field of view.

Much of the above consideration applies also to passive scanning, where this preobjective region is represented by *image space.* As shown in Figure 1.2, the off-axis flux arriving from a remote off-axis object point is directed on-axis toward the detector by the deflector in image space. Wide-angle performance remains required of the objective lens, as of a photographic lens collecting off-axis field flux.

1.5.1.2 Postobjective Scanning   Figure 1.10 illustrates a typical arrangement of postobjective scanning. Although it appears simply as a reversal of the scanner and lens positions of Figure 1.9, it exhibits

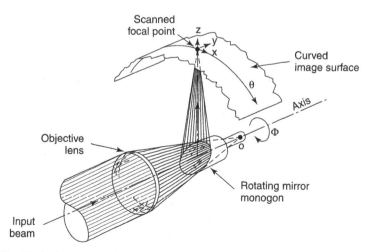

Fig. 1.10   Typical postobjective scanning by a monogon, generating a curved field. Shown in radial symmetry (with the illuminating beam focused on the axis), the scanned focal point locus is a portion of a perfect concentric circle.

very different performance. Most apparent for the active system is the generated curved field, imaged on a correspondingly curved image surface. When operated in perfect radial symmetry (introduced near the end of Section 1.5.1.1 and discussed in Section 3.4.1), this curve is perfectly circular, centered on the axis. This aids fundamentally in forming a scanned surface that allows simple fabrication for matching to high accuracy, where focus is critical (narrow depth of field). This is typical of the cylindrical surface in the internal drum scanner, as used in the graphic arts. Also, as identified above, when in radial symmetry, the image rotates about its projected axis. Although this is generally imperceptible when scanning a single isotropic spot (i.e., symmetric about its axis), the consequences of scanning nonisotropic spots (polarized, oblated, or multiple beam) and their derotation are considered in Section 4.3.5.5.

Another major distinction from preobjective scanning is the vastly reduced burden on the objective lens, for it now operates on-axis. The special complexity of design and fabrication of the flat-field lens in preobjective scanning (Sections 1.5.1.1, 4.3.5.2, and 4.6) is eliminated. Notably, in monochromatic illumination, the principal lens aberration of concern in postobjective scanning is spherical aberration—a parameter that may also be controlled elsewhere in the optical path.

In passive scanning, the postobjective region is identified as *object space*. This can extend over a substantial range; essentially the distance to the scanned object. In cases where this distance is so great that the flux arriving at the objective lens is nearly collimated, the depth of field is so great that it may not be perceived as curved. When, however, the object space is so shortened that field curvature is significant, then the object surface must comply with this curvature within tolerable focal depth, as for active scanning.

1.5.1.3 *Objective Scanning*   This third type of scanning, objective scanning, appears in several forms that exhibit common characteristics. Differing from the above-described techniques, which all depend on an *angular* deflection, objective scanning often imparts a transverse *translation* of the lens or of the scanned surface. Referring to Figure 1.1, and assuming for simplicity that the input beam to the lens is collimated, a transverse translation of the lens will also translate the image point P_i. This is analogous to translating the image field with respect to a fixed system of Figure 1.1 or to translating the detection system of Figure 1.2 with respect to a fixed object field. In these cases, the scanned point undergoes a linear shift that is uniquely not derived from an angular change of the beam entering the lens. (An angular change of the entire

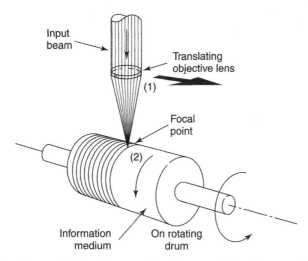

Fig. 1.11 Two forms of objective scanning for a rotating drum configuration. (1) The lens and its focal point translate with respect to the information medium, and (2) the information medium effectively translates with respect to the lens during drum rotation.

TABLE 1.2   Scanner-Objective Lens Characteristics

Type of Scanning	Ref. Fig. No.	Deflector Location	Scanning Process	Relative Lens Complexity	Allowed Scanned Function
Preobjective	1.8; 1.9	Before lens	Angular/fast	Complex	Typical flat field
Postobjective	1.10	After lens	Angular/fast	Simple	Curved field*
Objective	1.11	Lens or information surface	Translational/ relatively slow	Simple	Variable; flat; curved

* Perfectly circular when in radial symmetry.

assembly may be instituted, per Section 4.5.5, forming an arced objective scan.) A revealing illustration of two forms of linear objective scan appears in Figure 1.11, where one form is represented by translation of the lens and the other by the effective translation of the cylindrical surface under the lens. An information medium on the cylindrical surface, when unwrapped and flattened, exhibits a parallel line structure that appears as a raster. Another familiar raster form develops when the line scans are generated in direction x on a nominally flat information surface while this surface is translated in a perpendicular direction (y). In remote sensing, this surface translation effect is often achieved by the movement path of the vehicle carrying the sensing system, as discussed in Section 3.5.

1.5.1.4 *Summary of Scanner-Lens Relationships* The foregoing discussion of scanner-lens architecture is summarized in Table 1.2. Although the nomenclature applies to the active (e.g., flying spot) system (as represented in the referenced figures), all relationships apply, as well, to the passive system, in which the flux direction is reversed. Thus the first two entires in the Deflector Location column are represented by "image space" and "object space," respectively when considering passive systems (Fig. 1.2). Under Relative Lens Complexity, "complex" is due to wide field operation and "simple" is due to on-axis performance. It is also important to keep in mind the radial symmetry and image rotation characteristics that can become operative in postobjective scanning, detailed further in Sections 3.4.1 and 4.3.5.5.

CHAPTER 2

SCANNING THEORY AND PROCESSES

2.1 THE POINT SPREAD FUNCTION AND ITS CONVOLUTION

As Introduced in Chapter 1, scanning transforms a multidimensional function of spatial coordinates and time into a signal that is uniquely a function of time. Or it may perform the reverse transformation from a signal to an output spatial pattern. The first operation represents "reading" of a pattern to derive a signal for a data channel, and the reciprocal operation represents "writing," to form the output pattern from its corresponding signal. The electrical signal theory and the optical image theory are fundamentally analogous, motivating the worker who is trained in one field to appreciate the common relationships that unify both fields [John].

Establishing first the basic distinction between amplitude and intensity functions, Figure 2.1 illustrates one pair of functions of the variable x. The dashed curve forms the amplitude function $A(x)$ and the solid curve, the intensity function $I(x)$ such that

$$I(x) = \langle A(x,t)\, A^*(x,t) \rangle = |A(x)|^2 \qquad (2\text{-}1)$$

in which $*$ denotes the complex conjugate and the angular brackets represent the time average of the extremely high frequency (10^{14}–10^{15} Hz) of the electric vector of an optical electromagnetic field. The analogous equivalent to $A(x)$ is electrical voltage or current, which when squared to $I(x)$ is proportional to power and, over time, to energy. In optics, the

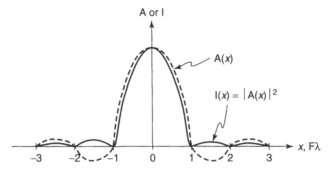

Fig. 2.1   Amplitude A(x), dashed curve, and intensity I(x), solid curve, of variable x, where I(x) = |A(x)|². Illustrating a typical PSF distribution (not to scale) derived from a uniformly illuminated rectangular aperture, where the nulls (1,2,3) are in units of Fλ. A(x) represents the coherent PSF and I(x) the incoherent PSF.

I(x) represents the optical lines of flux, which when incident upon an area represent flux density or power density. The response of all photosensitive materials and optical detectors, the eye included, is to the intensity function.

Figure 2.1 is recognized in the electrical domain as the "impulse response" of a linear system to a Dirac delta function (unit impulse) and in the optical field as the "point spread function" (PSF) image response to an ideal point object in a linear optical system. Because the PSF actually extends in two spatial dimensions (x and y), a more direct analogy to the one-dimensional signal impulse response is the one-dimensional "line spread function" image of an ideal line object [D&S].

2.1.1   PSF Developed from Uniform Illumination of an Aperture

Observation of Figure 2.1 reveals that the nulls in intensity (or zero-amplitude crossovers) appear at equispaced intervals +1, 2, 3, ... (in units of Fλ; F $\equiv$ f-number of the beam converging to the PSF, and $\lambda \equiv$ wavelength). This is representative of a particular PSF, that derived from filling a rectangular aperture (or pupil) with light of uniform intensity, forming in the far field* a function similar to that illustrated in Figure 2.1 and expressed by

---

* Far field $\equiv Z_f$, the distance from the aperture along the z-axis of value $Z_f > 50D^2/\lambda$. Near field $\equiv Z_n < D^2/\lambda \simeq$ Rayleigh range. Intermediate distances are transition region.

$$A(x) = \frac{\sin \pi x}{\pi x} \equiv \mathrm{sinc}(x) \qquad (2\text{-}2a)$$

and

$$I(x) = \left[\frac{\sin \pi x}{\pi x} \right]^2 \equiv \mathrm{sinc}^2(x) \qquad (2\text{-}2b)$$

The significance and application of this uniformly illuminated rectangular aperture is discussed in Section 2.4. Except for the case of the infinitesimally narrow x and infinitely long y slit (analogous to the signal delta function), the rectangular aperture is two-dimensional, forming in the far field (or in a focal plane per Section 2.4.1) a PSF that is represented more completely as

$$A_\square(x, y) = \mathrm{sinc}(x)\mathrm{sinc}(y) \qquad (2\text{-}3a)$$

and

$$I_\square(x, y) = [\mathrm{sinc}(x)\mathrm{sinc}(y)]^2 \qquad (2\text{-}3b)$$

A point spread function that, in cross section, appears very similar to that of Figure 2.1 is that derived from filling a round aperture (or pupil) of radius r with light of uniform flux density. The resulting focused PSF intensity distribution $I_o(r)$ is that of the familiar Airy disc, represented by Equation 2-4b

$$A_\circ(r) = \frac{2J_1(r)}{r} \qquad (2\text{-}4a)$$

and

$$I_\circ(r) = \left[\frac{2J_1(r)}{r} \right]^2 \qquad (2\text{-}4b)$$

in which $J_1(r)$ is the Bessel function of the first kind of order 1 of the variable r. In contrast to the separable quadrature functions of the PSF that develop from a rectangular aperture, Equations 2-3a,b, this PSF exhibits rotational symmetry (when derived from a round aperture).

A significant feature of this PSF is that the first null in intensity occurs at $r = 1.22\,\mathrm{F}\lambda$, as distinguished from the above first null appearing at $x = 1.0 F\lambda$. Thus this round PSF is some 22% larger in diameter than the width of the PSF from a rectangular aperture. This feature is developed further as the aperture shape factor a in the discussion of spot size and scanned resolution below (Section 3.2). Although both functions exhibit decreasing intensity of the successive lobes, it is noteworthy that successive nulls of this function occur at progressively smaller intervals of r, compared to the equal intervals of null of the sinc^2 function. Also distinguishing, is that the first two side-lobe intensities of the Airy function are approximately 1/3 the magnitude of the corresponding lobes of the sinc^2 function. Figure 2.1 is representational and is not to scale. Scaled plots of the two functions appear elsewhere in the literature [Mar].

2.1.2 PSF Developed from Aperture Illumination with a Gaussian Distribution

The point spread functions discussed above develop from illumination with *uniform intensity distributions* of two prominent aperture shapes (rectangular and round/elliptic). Other aperture shapes that may be so illuminated are discussed in Section 3.2.1.1.

A prominent illumination function that is uniquely nonuniform in intensity is that of the Gaussian distribution. Known as the fundamental or TEM_{00} resonator mode [Len], this develops as a consequence of the high multiplicity of self-reproducing cyclic wave fronts in a well-aligned laser resonator. It is typical of the radiation from most well-behaved lasers. Illustrated in Figure 2.2, several of its irradiance parameters are identified. Notably, there is no distinctive feature for identification of its "width" or "diameter." It exhibits no nulls, but rather a monotonic decrease in the intensity of its "skirts" at increased width.

Figure 2.2 reveals several key markers that identify this distribution. Given the equation of the Gaussian intensity function

$$I(r) = I(0) \exp\left[-2(r/w_o)^2\right] \tag{2-5}$$

where $I(0)$ is the maximum value at its radial center and w_o is the beam radius at which its intensity is $1/e^2 \approx 0.135$ of its central maximum value. $I(r)$ is its value at radial distance r. This Gaussian intensity equation is also a squared function of the amplitude $A(r)$ (of, e.g., the electric field). Thus $A(r) = A(0) \exp-(r/w_o)^2$ is the square root of Equation 2-5.

At radial distance $r = w_o$, the amplitude is $1/e$ of the central peak value and the intensity is $1/e^2$ of the corresponding maximum intensity.

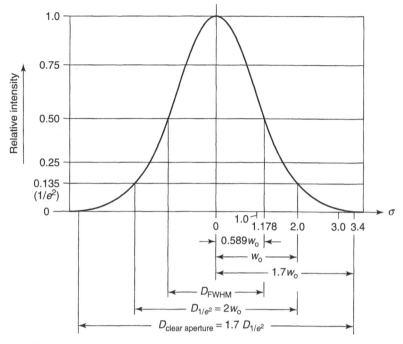

Fig. 2.2   The Gaussian intensity function, showing several identifying characteristics. Most significant is its standardized radius w_o set at the 2σ point of its normal distribution, at which the intensity measures $1/e^2$ of its peak value. The corresponding diameter is $D_{1/e^2} = 2w_o$, and its diameter at $\frac{1}{2}$ intensity (FWHM) is $0.589\,D_{1/e^2}$. To transfer the full value of the Gaussian requires a clear aperture, indicated at a diameter of $1.7\,D_{1/e^2}$.

This radial distance is identified as the 2σ point in the normal or probability distribution—spaced two standard deviations on either side of the central maximum value. Discussed further under the topic of scanned resolution (Chapter 3), this radial distance $r = w_o = \pm 2\sigma$ is the most utilized identification of the width of the Gaussian distribution. This applies not only to the relatively wide beam width $D_{1/e^2} = 2w_o$ (Fig. 2.2) illuminating a focusing lens but also to its narrow focused "spot"; the PSF. Contrary to the shape transformations experienced by the uniformly illuminated functions discussed above, this Gaussian beam shape is sustained as it propagates through linear optics, such that "once a Gaussian, always a Gaussian"—from the near field through to the far field. To maintain this characteristic, it is necessary to transfer essentially the full Gaussian function through an aperture of minimum width represented in Figure 2.2 as

$$D_{\text{clear aperture}} = 1.7\,D_{1/e^2} \qquad\qquad (2\text{-}6)$$

If the beam is truncated below 1.7x, its PSF expands and minor side lobes appear.

All the discussed PSFs exhibit a common appearance in their principal lobes. If higher-order lobes exist, they are of monotonically decreasing intensity, as are the tapering skirts of the Gaussian function. Thus we consider that operational PSFs are dominated by a principal lobe that is similar to that of the Gaussian function within its $1/e^2$ values. Following the central limit theorem [Lev], this becomes further cogent with the inclusion of any pseudo-randomness in a real system due to, for example, aberration or other temporal or spatial perturbation.

Identified in Figure 2.2 is another significant characteristic of the Gaussian contour, its radius or diameter at which the intensity equals $\frac{1}{2}$ of its peak value. A popular descriptor for this beam diameter is full width at half maximum (FWHM). The relationship between these two prominent identifiers of the Gaussian beam, at $1/e^2$ intensity (D_{1/e^2}) and at $\frac{1}{2}$ intensity (D_{FWHM}), as determined from Equation 2-5, is

$$D_{FWHM} = 0.589 D_{1/e^2} \qquad (2\text{-}7)$$

A notation that relates to scanned resolution and the developed spot size is the aperture shape factor a, discussed in Chapter 3. Its value is an adjustment factor for the PSF or spot size, accomodating and normalizing various aperture illumination conditions. For the focused Gaussian spot intensity distribution, following Equation 2-7,

$$a_{FWHM} = 0.589\, a_{1/e^2} \qquad (2\text{-}8)$$

Because the value of a_{1/e^2} for the untruncated Gaussian is shown in Section 2.3.2 as

$$a_{1/e^2} = 4/\pi \simeq 1.27 \qquad (2\text{-}9)$$

then, per Equation 2-8,

$$a_{FWHM} = 0.589(4/\pi) \simeq 0.75 \qquad (2\text{-}10)$$

Chapter 3 discusses these and other parameters relating to scanned resolution.

2.1.3 Scanning—Controlled Movement of the PSF; Its Convolution

The mission of a scanning process is to render controlled movement of a point spread function—often described as the focal "spot" or "dot." This "sampling aperture" serves as a tiny "window" through which flux is either sampled from or deposited on an information medium. The system that provides this movement is the scanner, along with its associated optics and positional control elements, as discussed in Chapters 4 and 5. Here we are primarily concerned with the informational parameters of scanning and their effects on the integrity and fidelity of the transfer process.

With the PSF identified as $I(x)$, Figure 2.3 represents its most prevalent type of scanning—linear uniform translation. Utilizing the earlier precept of the commonality of the principal lobe of the PSF, the illustrated $I(x)$ becomes generic. Its influence in a subsequent scanning process is further standardized by normalizing the complete $I(x)$ to unity, that is, $\int I(x)\mathrm{d}x = 1$. Assuming, first, that $I(x)$ is a stationary function, that is, that its local shape is independent of its position [D&S], Figure 2.3 shows the PSF performing a "sliding" operation at a uniform velocity v over a time t, covering the distance $x_v = vt$. This variable position of the PSF is represented by $I(x_v - x)$, the new center about which resides $I(x)$. With the ideal PSF assumed symmetric about its central maximum (as is the symmetry of the Fourier transform), then $I(+x)$ and $I(-x)$ are identical, forming the shifting function identified as $I(x_v - x)$.

One system of interest is, for example, the generation of a "continuous tone" facsimile copy of an original document. This is conducted by

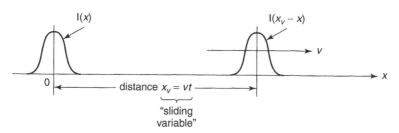

Fig. 2.3 Illustration of scanning with a generic intensity function $I(x)$, executing a sliding operation at velocity v over distance x_v during time t. Even if conducted with a coherent optical beam, the process of scanning across different regions at different times destroys the deposited or measured coherence, forming thereby the indicated incoherent PSF, $I(x)$.

scanning a PSF across a surface of varying reflectance, f(x), deriving thereby the signal g(t) that then modulates and exposes a photosensitive material point by point synchronously to reconstruct a "duplicate" spatial pattern g(x) to some quality expectation. Addressing first the one-dimensional case and assuming perfect electrical-to-optical transfer for simplicity, the output function g(x) is equal to the input function f(x) convoluted by the shifting function. This is represented [Hob] by the convolution integral

$$g(x) = \int f(x) I(x_v - x) dx \qquad (2\text{-}11)$$

which is illustrated in Figure 2.4a. The input function f(x) represents an arbitrary intensity contour along distance x. A PSF of full relative intensity I(x) translates at constant velocity v to measure f(x) at all points along its continuous path. Starting at $x = 0$ (and continuing through x_a, x_b, x_c, and x_d), the measurements are the products of each area under

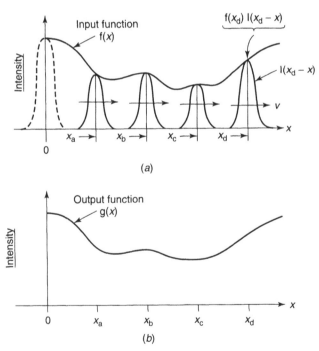

Fig. 2.4   (a) Input function f(x) of arbitrary intensity variations scanned by a PSF of finite width I(x). Notation follows that of convolution integral, Equation 2-11. (b) Resulting output function g(x) shows some loss of detail due to scanning by a finite-width PSF; notably in vicinity of x_c.

the overlapping curves of I(x) and f(x). I(x) takes a sequence of dx spacings, until the entire function f(x) is measured. Because this PSF exhibits a finite width, each measurement is smoothed by the overlapping components of the functions. If the PSF were an ideal point object delta function, then each measurement would provide for rigorous reconstruction. As illustrated, this relatively broad PSF tends to overlap and thereby smooth the fluctuations in the original f(x) of Figure 2.4a, notably around x_c, as manifest in the loss of some corresponding fine detail in the reconstructed g(x) of Figure 2.4b. In this respect, the scanning function has the further characteristic of acting as an electrical low-pass filter.

An objective of this pursuit is to predict the loss in quality of the reproduction, based on the characteristics of the scanning "spot." One limit case is heuristically apparent, that a pure delta function PSF imposes no loss. Clearly, the narrower the PSF—that is, the smaller the spot relative to the breadth of the fluctuation in f(x), the less is the deterioration. The discussion of the modulation transfer function (MTF) in Section 2.4.2 will contribute to this prediction of scanned quality.

If we are considering an image area, the simplified one-dimensional case represented in Equation 2-11 becomes ultimately a two-dimensional convolution, whether by continuous or discontinuous sampling. Assuming equal continuity in both x and y directions, as would be a typical photographic spread function convolving a two-dimensional image [D&S], this is represented more rigorously by the two-dimensional convolution, in which the integration is limited to the image area rather than covering all space,

$$g(x,y) = \iint_{\text{image}} f(x,y)\, I(x_{vx} - x, y_{vy} - y)\, dx, dy \qquad (2\text{-}12)$$

2.2 QUANTIZED OR DIGITIZED SCAN

The most prevalent form of piecewise scanning, either for capture or for display, is by the formation of a rectilinear raster [W&Z], as represented in Figure 2.5. Other patterns may be deployed [W&Z], but they entail more complex formation and analysis and less homogenous image coverage. The television image, notably as viewed on a "black-and-white" cathode ray tube (CRT) having a continuous phosphor coating [Sherr], is a familiar rectilinear raster image. The x-directed scan impinges a fine beam of electrons on the phosphor screen in a predetermined relation to the desired intensity of the resulting phosphorescent spot. This "paints" a single line of the intensity information

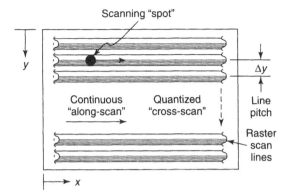

Fig. 2.5 Formation of a raster with a scanning "spot." The x-directed along-scan lines are formed continuously and spaced apart by a line pitch Δy, shown exaggerated. This cross-scan periodicity imposes quantizing or sampling of the information in that direction. Valid reconstruction requires an adequate number of scan lines subtending the shortest y-directed information cycle.

"along-scan." The two-dimensional area is covered by incremental displacement of the scanning spot "cross-scan," forming thereby additional lines sequentially into a raster "frame." Although the along-scan line may indeed be represented by Equation 2-11, the successive lines no longer derive from infinitesmal cross-scan increments of dy. Rather, they occur displaced by a "line pitch" Δy. This space must be established in the original scanning process to be sufficiently small, and be reconstructed accurately, to convey the y information rigorously.

In Figure 2.5, the scan lines are rendered as strips formed by a Gaussian-like "spot," and their pitch Δy is exaggerated to show the structure. With the typical line pitch $\Delta y \leq$ PSF, Δy would then be shown at about $\frac{1}{2}$ of its illustrated value. Another characteristic of Figure 2.5 is that the along-scan components are assumed derived from continuous analog scan of the original and reproduced information (as with a continuous flying spot scanner). With motivation in some applications to digitize the original and reconstructed along-scan information, and with the inevitable quantized Δy spacing in cross-scan within a raster, the criteria for such digitizing and line spacing merit attention.

2.2.1 The Sampling Criterion

Consider Figure 2.6, which shows a portion of a sinusoidal intensity function having its highest cyclic frequency $\tilde{x}_{max}$, corresponding to the shortest cyclic period of $1/\tilde{x}_{max}$. This could, for example, represent the

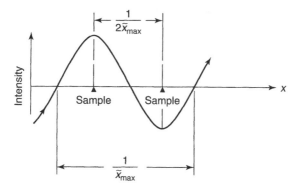

Fig. 2.6 Assumed sinusoidal signal (or intensity) fluctuation of highest informational cyclic frequency x_{max}. Nyquist sampling criterion requires a sampling rate of $f_{max} \geq 2x_{max}$, at least two samples per information cycle for valid signal reconstruction. Sampling is shown phased at the peaks and troughs of the signal and spaced at the maximum criterion of $1/2x_{max}$.

fluctuating component in f(x) of Figure 2.4a, notably, the "fastest" variation, in the vicinity of x_c. Given such an arbitrary signal component and the desire to sample it at a minimum rate (for economy) so that it may still be reconstructed validly, we apply the Nyquist sampling criterion [BTL, D&S]. This recognized criterion requires that the fluctuating component $\tilde{x}_{max}$ be sampled at a rate f_s such that its frequency be at least twice that of $\tilde{x}_{max}$. That is,

$$f_s \geq 2\tilde{x}_{max} \qquad (2\text{-}13)$$

The frequency may be either temporal or spatial, at a rate of at least two samples per information cycle. With appropriate signal handling and low-pass filtering to cut off effectively at $\tilde{x}_{max}$ and with adequate signal-to-noise ratio to allow high amplification in regions of low signal, the original function may be reconstructed theoretically perfectly with as few as two samples per information cycle. In practice, to overcome the imperfections of noninfinitesimal sampling function (PSF), data handling and band-limiting factors, and sampling at inappropriate phase [L&U] within the cyclic fluctuation, oversampling is often applied.

The problem of sampling at inappropriate phase may be seen in Figure 2.7, which shows the sampling of a signal such as that in Figure 2.6 under two conditions (a) and (b). The upper part of condition (a) is essentially a repeat of Figure 2.6 (solid curve) carried over a few cycles. Sampling (S) is at exactly two per cycle, occurring at the peaks

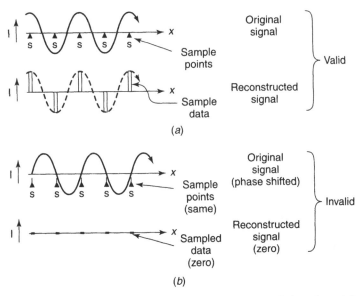

Fig. 2.7 Signal components of variable phase that are sampled at the minimum criterion of $f_s = 2x_{max}$ under two conditions of relative phase. Condition (a) shows sampling at the same peak-and-trough phase of Fig. 2.6, allowing a valid reconstruction after bandlimiting and amplification. Condition (b), sampling the 90° out-of-phase signal at the identical periodicity, yields a fundamental loss of information that frustrates attempts at reconstruction, demonstrating the need for oversampling randomly phased data.

and troughs (maxima and minima) of the original signal. Just below are shown the sampled data and the reconstructed signal (dashed curve) after low-pass filtering and amplification. It forms, validly, an essentially perfect reproduction of the original signal. Condition (b) shows the identical original signal (upper solid curve), which, sampled (S) *at the same two per cycle*, occurs now 90° out of phase—at the zero crossovers. The reconstructed signal reveals a loss of information that resists any practical attempts toward reconstitution—simply invalid sampling at this phase. Given an arbitrary signal that is sampled at an independent regular clock rate, this very probable impasse is overcome by oversampling.

One of the more familiar examples of such oversampling is that conducted in current NTSC (National Television Standards Committee) practice here [BTL] and in related international standard practice. The total number of 525 lines (NTSC) in the vertical direction reduces to 488 active lines because of the 7% of the vertical blanking period allotted to the vertical retrace interval. Most significantly regarding our interest, early (1934) human factors testing revealed that, to overcome much of the sampling loss described above, the remaining 488 active scan lines must represent no more than 0.7 of that number of data points, that is, 342 half-cycles of reliably retrievable image data. This factor of 0.7 became standardized throughout the world, identified as the Kell factor after one of the prominent early researchers [Kell]. In the context of our earlier discussion, application of a Kell factor of 0.7 represents data sampling at approximately three samples per information cycle.

An appropriate follow-up question is, At what rate of sampling can we expect the reconstructed signal (devoid of other quality perturbations) to have recovered essentially all of its original content? Or, how much need one oversample arbitrary image fluctuations to allow effective reconstitution of the original image? This question was addressed in the mid-1960s [CBS], during the research phase of phototransmission system development. It was concluded that providing four samples per shortest information cycle—or $2x$ oversampling—renders an almost fully restored image. Validation appears in a 1962 review paper [Lew] and, more recently, for pixelated image systems [H&Bo]. Substansive discussion of oversampling appears in the audio digitization literature [Poh].

2.3 GAUSSIAN BEAM PROPAGATION

Section 2.1.2 introduced the Gaussian beam intensity distribution and some of its identifying features, to quantify this function, that is most often scanned. Its geometric and radiant characteristics along the beam propagation path, discussed here, are of equal significance, for they denote its transport in homogenous space and as modified by intervening optical elements.

2.3.1 Representation and Development of the Gaussian Beam

An illustration of the Gaussian beam viewed transverse to its propagating Z-axis appears in Figure 2.8. The solid line curved portions, symmetric about the Z-axis, represent the boundary of the $1/e^2$ intensity value. Any section perpendicular to the Z-axis exhibits an irradiance represented in Figure 2.2. Figure 2.8 shows a beam "waist" of radius w_o (or diameter d_o) that can be the "spot size" of beam focusing. Noting the dotted waist tangent lines parallel to the Z-axis, the position of minimum beam diameter may also represent the output port of a laser

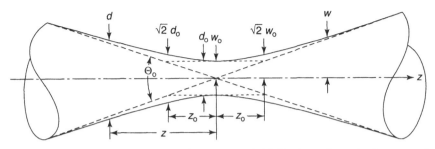

Fig. 2.8   Gaussian beam (solid lines) converges initially through angle Θ_0 to a waist radius w_0 (diameter d_0) and then diverges again through angle Θ_0 as it propagates in the Z-direction. The Rayleigh range is identified as $\pm z_0$ at which distance the beam radius (and diameter) increase by a factor of $\sqrt{2}$. General beam diameter d is located a general distance z from beam waist.

that is radiating an initially "collimated" beam of diameter d_0. Although this figure may appear to represent rotational symmetry, the illustration and the following discussion apply equally along the major and minor axes of a beam of elliptic cross section.

Additional notation on Figure 2.8 includes the axial distance z_0, a parameter described as the Rayleigh range—the distance from the beam waist where its subtense is increased by a factor of $\sqrt{2}$. This is noted as the radius $\sqrt{2}w_0$ and the diameter $\sqrt{2}d_0$. A general beam radius is identified as w, and a general beam diameter is indicated as d, located at a general distance z from the beam waist.

It is useful to appreciate the geometric significance of this representation, and hence its extension to optics. The function illustrated in Figure 2.8 may be recognized as that of a hyperbola of eccentricity e > 1 [Bur], centered on its axes. With substitution of the above nomenclature, this general equation reveals directly,

$$\frac{w^2}{w_0^2} \equiv \frac{d^2}{d_0^2} = 1 + \frac{z^2}{z_0^2} \qquad (2\text{-}14)$$

which is rarely expressed in that simple form [S&T]. A one-step variation [Osh] appears as

$$\frac{d^2}{d_0^2} = 1 + \frac{\Theta_0^2 z^2}{d_0^2} \qquad (2\text{-}15)$$

in which $\Theta_0 = d_0/z_0$, the (converging or diverging) beam angle.

In Figure 2.8, the solid curve is the hyperbola and the crossed dashed lines, forming the angle Θ_o, are its asymptotes. In geometric optics, absent diffraction, a converging beam focuses to a theoretical point: the crossover of the asymptotes. The diffraction angle Θ_d is determined by the wavelength λ, the subtense d of the radiating aperture, and a coefficient identified as aperture shape factor a (Sections 2.1.2 and 3.2). Θ_d is given as

$$\sin \Theta_d = a\frac{\lambda}{d} \qquad (2\text{-}16)$$

($a = 4/\pi$ for the Gaussian beam, with d measured across its $1/e^2$ intensity points). When d is at a large distance z before focus (Fig. 2.8), this diffractive spread Θ_d forms the finite beam waist d_o, where it would otherwise collapse to a point. Identifying in Figure 2.9 this distance z as an effective focal length f from reference diameter D, and the focused subtense d_o as the "spot size" of notation δ in this work, then (for $\sin \Theta_d \simeq \Theta_d$), the spot size is

$$\delta = \Theta_d f = a\left(\frac{f}{D}\right)\lambda = a F \lambda \qquad (2\text{-}17)$$

where $F = f/D$ is the f-number of the converging beam, as detailed in Section 2.3.2.

A revealing alternative derivation of the hyperbolic function of Figure 2.8 identifies d_o as the width of a *geometric* collimated beam

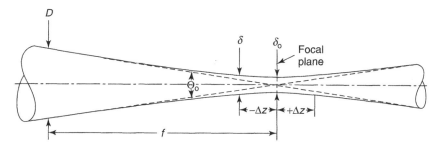

Fig. 2.9 A variation of Fig. 2.8, illustrating a Gaussian beam converging to a focal plane. The waist diameter is now the "focal point" of "spot size" δ_o, located a distance f from a remote beam diameter D. This establishes the beam f-number $F = f/D$. The focal depth is indicated as $\pm\Delta z$, which determines the tolerable growth of the spot size δ compared to δ_o.

(dotted tangent lines extended parallel to z) that is not correlated with a *diffractive* width component $d_z = \Theta_o z$ spreading beyond d_o over distance z. Assuming a root-sum-square (RSS) composite uncorrelated addition of the fixed width d_o and the expanding width d_z [B&S], one may write immediately

$$d = \left(d_o^2 + \Theta_o^2 z^2\right)^{1/2} \qquad (2\text{-}18)$$

This equation is a direct restatement of Equation 2-15.

A more common, though more elaborate, form of the Gaussian beam equation utilizes the Rayleigh range,

$$z_o = d_o / \Theta_o \qquad (2\text{-}19)$$

defined earlier, and as deduced from Equations 2-14 and 2-15. As in Equation 2-16, the diffractive spread Θ_o ($\approx \sin\Theta_o$) is represented numerically by

$$\Theta_o = \left(\frac{4}{\pi}\right)\frac{\lambda}{d_o} \qquad (2\text{-}20)$$

forming the Rayleigh range

$$z_o = \frac{\pi d_o^2}{4\lambda} \qquad (2\text{-}21)$$

or

$$z_o = \frac{\pi w_o^2}{\lambda} \qquad (2\text{-}22)$$

Substituting Equation 2-22 into Equation 2-14 yields the customary representation of Gaussian beam propagation,

$$\frac{w}{w_o} = \left[1 + \left(\frac{\lambda z}{\pi w_o^2}\right)^2\right]^{1/2} \qquad (2\text{-}23)$$

whereas the simpler Equations 2-14 and 2-15 allow perception of the fundamental formation of Equation 2-23.

2.3.2   Gaussian Beam Focusing Characteristics

The discussion surrounding Equations 2-16 and 2-17 introduced the focusing properties of the Gaussian beam, in which the f-number F and its related parameters are prominent factors. Given a converging beam, as shown in Figure 2.9, its $1/e^2$ width D is located a distance f from the focal plane. No longer the general beam dimension d of Equation 2-17, this D often denotes the beam width, which, when converged over the distance f, develops the focused spot size δ_o. Also, when positioned at the fulcrum of an angularly scanned beam, the D is a fundamental parameter that determines *scanned resolution*, detailed in Chapter 3. Unlike the f-number rating of a lens, whose diameter may be significantly larger than the beam diameter D, this f-number F describes *the converging cone* of a focusing beam. It is ascribed *to the beam*, rather than to the lens. Defined similarly as

$$F \equiv \frac{f}{D} \qquad (2\text{-}24)$$

this F is established at a beam location that is sufficiently remote from focus (evaluated in Section 2.3.2.1), where its width D is effectively dominated by the converging geometric asymptotes. It is a constant of the beam.

The focused spot size δ_o, following Equation (2-17), is represented by

$$\delta_o = a F \lambda \qquad (2\text{-}25)$$

in which a is an aperture shape factor and λ is the wavelength. When D and δ_o are both measured across their $1/e^2$ intensity points (on the converging beam and at the focused "waist," respectively) and the beam is unapodized to the extent that Equation 2-6 is satisfied, then the Gaussian aperture shape factor is given by

$$a_g = \frac{4}{\pi} \simeq 1.27 \qquad (2\text{-}26)$$

This derives from the diffraction field of the Gaussian beam, as by Equation 2-20.

For application to Gaussian beam focusing and depth-of-focus tolerancing, a prominent variation to the Rayleigh range of Equations

2-19 and 2-22 is obtained by substituting the spot size δ of Equations 2-25 and 2-26 into Equation 2-21. With $d_o \equiv \delta$, this yields the Rayleigh range

$$z_o = \frac{4}{\pi} F^2 \lambda \qquad (2\text{-}27)$$

2.3.2.1 A Criterion for Determining F

The f-number $F = f/D$ measures the beam diameter D at a sufficient distance f from focus. As noted after Equation 2-24, we seek a particular distance $z \equiv f$ at which the corresponding beam diameter $d \equiv D$ approaches that of the expanding asymptotes in Figure 2.8.

Per Equation 2-18, the composite Gaussian function is represented by

$$D^2 = d_o^2 + \Theta_o^2 f^2 \qquad (2\text{-}28)$$

and the equation of the asymptotes alone is

$$d_z = \Theta_o f \qquad (2\text{-}29)$$

Dividing Equation 2-28 by the square of Equation 2-29 and using Equation 2-19, one obtains

$$\frac{z_o^2}{f^2} = \frac{D^2}{d_z^2} - 1 \qquad (2\text{-}30)$$

When the ratio $D/d_z \rightarrow 1$, this denotes the proximity to dominance of D by d_z, and the ratio f/z_o determines the distance f in multiples of z_o. Letting $D/d_z = 1.01$ (equal within 1%), one finds that for secure measurement of the f-number, that

$$f \geq 7z_o \qquad (2\text{-}31)$$

That is, valid measurement of f and D is obtained when f exceeds 7 Rayleigh ranges.

2.3.2.2 Evaluation of Depth of Focus

An immediate consequence of defocus of the Gaussian beam is obtained from Equation 2-14. Let the defocus distance z correspond to one Rayleigh distance z_o. With $z/z_o = 1$, then Equation 2-14 shows $d/d_o \equiv \delta/\delta_o = \sqrt{2}$. That is, with

TABLE 2.1   Values of Parameter c for Spot Size Increase due to Defocus of a Gaussian Beam by $\Delta z = \pm cF^2\lambda$

% Spot Size Increase	Corresponding δ/δ_0	c
41.4	$\sqrt{2}$	$4/\pi = 1.273$
30.0	1.3	1.06
20.0	1.2	0.85
10.0	1.1	0.58
5.0	1.05	0.41
0	1.0	0

defocus equal to the Rayleigh range, the spot size δ is enlarged by over 41%.

The term $F^2\lambda$ in Equation 2-27 appears in defocus equations other than those of Gaussian beams. Because the value $4/\pi$ in Equation 2-27 applied only to the Rayleigh range, it is expedient to express a general defocus range with a coefficient c instead of the fixed $4/\pi$. Then the expression

$$\Delta z = \pm cF^2\lambda \qquad (2\text{-}32)$$

represents the focal shift Δz that causes a known spot size increase. Table 2.1 expresses these values, derived from Equations 2-14 and 2-17. This data is significant for scanner designs in which the Rayleigh criterion of $\sqrt{2}$ (41.4% increase) exceeds system objectives for tolerable spot growth. For example, if no more than 20% spot size increase is allowed, Δz may not exceed the value resulting from c = 0.85 in Equation 2-32. This constrains the tolerable Δz to $0.85 \div (4/\pi) = 0.67$ of the Rayleigh range.

2.4   SCANNED QUALITY CRITERIA AND THE MODULATION TRANSFER FUNCTION

In the prior discussion, a few scanned quality criteria were identified, such as the effects of the spot size relative to the narrowest desired signal fluctuation (Fig. 2.4), the sampling rates relative to this narrowest signal variation (Section 2.2.1), and the tolerable departure from perfect focus (Section 2.3.2.2). All of these factors—and more—such as the effects of aberration and apodization and the effects of tandem compounding of subsystems may be quantified quite rigorously with application of the modulation transfer function (MTF). Although a comprehensive discussion of this multidisciplined topic is beyond the scope of this work, it is presented here with sufficient heuristic orientation for understanding of its analytic power in the field of optical scanning.

2.4.1 The Fourier Transform

The operation that forms the MTF is the Fourier transform [Bro, D&S, P&T, Shu]. In the field of optics and image handling, Fourier transformation relates the *spatial orientation* f(x) of an object to the distribution of its *spatial frequency* components F($\tilde{x}$)—their *rates of change* in units of, say, cycles per millimeter. In the analogous electrical signal domain, it relates the temporal variations in signal content f(t) to the distribution of its frequency components f(v) in units of, say, cycles per second. In this context, the Fourier transform functions as a form of spectrum analyzer.

Consider Figure 2.10, which shows (solid line) the signal intensity variation as detected from a single scan of an input object (as scanned with a PSF that is much narrower than a half-cycle of its highest-frequency component). This line is composed of a range of intensities and rates of assumed sinusoidal fluctuation: from a slowest "low-frequency component," to a moderate "midfrequency component," to its fastest (although weakest) variation marked "high-frequency component." Although a generic input object is likely to be composed of a continuum of cyclic rates, only three are identified here. The purpose of this "pre-Fourier transform" exercise is to imagine plotting, per Figure 2.11, the relative magnitudes of the full range of measured sinusoidal spatial frequencies—an extremely tedious real task. When realized, it shows

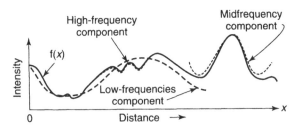

Fig. 2.10 Signal intensity variation f(x) as detected from single line scan of input object. Identifies assumed sinusoidal fluctuating components of signal having low-, mid-, and high spatial frequencies.

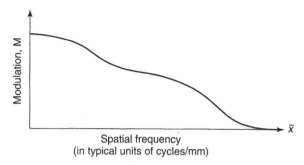

Fig. 2.11 Modulation spectrum of input signal frequencies. Shows assumed distribution of intensities of sinusoidal fluctuating components, as derived from a full range of spatial frequency components per Fig. 2.10.

the distribution of depths of modulation of the various assumed sinusoidal spatial frequency components that form the image. This is analogous to the *power spectrum* of the signal.

As is conventionally represented, modulation depth is taken as

$$M = \frac{I_{max} - I_{min}}{I_{max} + I_{min}} \qquad (2\text{-}33)$$

in which I_{max} and I_{min} are the magnitudes at the peak and trough, respectively, of the sinusoidal swing. Thus, M is the ratio of the actual peak-to-peak swing to the potential full excursion. All the modulated components are assumed to be composed of elementary sinusoidal waves, as are the elements of a Fourier series [Shu].

The resulting distribution of input frequencies illustrated by Figure 2.11 may be identified as an "input modulation transfer function." Rigorously, the modulation transfer function descriptor is reserved to represent the Fourier transform of the point-spread or line-spread functions discussed below. Highlighted here is the important concept that the above-hypothesized laborious task of developing the input spectrum distribution may be accomplished much more elegantly and directly by taking the Fourier transform of the input intensify function. The Fourier transform $F(\tilde{x})$ of a function $f(x)$ and its inverse relationship (identified as a Fourier transform pair) are represented by

$$F(\tilde{x}) = \int_{-\infty}^{\infty} f(x)e^{-i2\pi\tilde{x}x}dx \qquad (2\text{-}34)$$

and

$$f(x) = \int_{-\infty}^{\infty} F(\tilde{x}) e^{i2\pi\tilde{x}x} d\tilde{x} \qquad\qquad (2\text{-}35)$$

The $2\pi\tilde{x}$ may be represented by the radian frequency ω. Also, the integrals may appear preceded by normalizing terms of a $1/2\pi$ form [D&S, P&T]. These complex equations are composed of real and imaginary parts. Unless indicated otherwise, subsequent discussion relates to the modulus term. The phase term will be zero when, as is most prevalent, the spread function is assumed to be both real and even.

Whereas the Fourier series (summation) reveals the composition of a periodic function as a superposition of harmonic multiples of a fundamental frequency, the Fourier transform performs a continuous integration of a nonperiodic function, yielding a continuum of sinusoidal spectra. It is informative to view some related examples of forward and inverse transformations [D&S, P&T].

The Fourier transform serves in optical image formation in another fundamental manner. Section 2.1 introduced the development of the PSF as derived from different aperture shapes, describing their appearance at the "far field" focal plane. More explicitly, an illuminating beam having a given amplitude distribution in its near field exhibits a unique diffractive amplitude distribution in its far field. When this incident beam propagates through the aperture of a positive lens, the focusing process of the lens forms the diffracted far field amplitude distribution whose intensity distribution is the PSF, effectively brought forward to the focal plane. This fundamental process may be represented analytically as the |Fourier transform|^2 of the incident amplitude distribution at the aperture.

2.4.2 The Modulation Transfer Function

One may perform the next step in quality analysis by taking the Fourier transform of this PSF to yield its modulation transfer function (MTF) and its phase transform function (PTF), if it exists. Also, the operational equivalent to Fourier transforming a Fourier transform is convolving the aperture function with itself (rigorously, with its complex conjugate), thereby yielding the MTF directly from the aperture function.

These procedures are represented graphically in Figure 2.12. Assumed for simplicity is a rectangular aperture (upper left corner) of interest in only one direction ξ illuminated with a complex amplitude function $A(\xi)$. Just below appears its Fourier transform $A(x)$, the amplitude distribution in the focal plane, the sinc function of Equation 2-2a. Squaring this amplitude function yields its intensity distribution

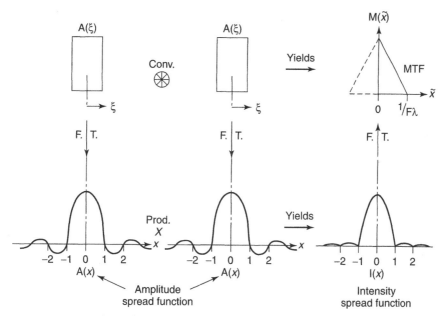

Fig. 2.12   Alternate analytic routes to the MTF. Top row: convolution of the A(ξ) complex amplitude aperture functions (autocorrelation) yields the MTF. Bottom row: Product of the Fourier transforms of the amplitude aperture function is the intensity spread function. Its Fourier transform is the MTF.

(bottom right), proportional to Equation 2-2b. This forms the PSF. Finally, moving upward at the right, the Fourier transform of this PSF yields the MTF of a rectangular aperture, illustrated as the solid-line right side of the symmetric triangle. Thus, starting with a uniform rectangular aperture (upper left in Fig. 2.12 and following counterclockwise), it is determined that the MTF of this aperture is represented by a triangular function.

Because the aperture distribution itself ulitmately determines the PSF, one can go directly from this pupil function A(ξ) to the MTF. This is achieved by convolving (rolling wheel symbol) the aperture with itself (complex conjugate); this is termed autocorrelation. Significant MTF data have been complied by using this elegant equivalent geometric process, as represented in a subsequent discussion (Section 3.2.1.1).

In determining scanning system quality, we are interested in the effect of the scanning spot upon the integrity of the resulting scanned output. One method of evaluating this is discussed in Section 2.1.3, where per Figure 2.4 we convolve the input function with the scanning PSF. This convolution is reillustrated in the upper half of Figure 2.13,

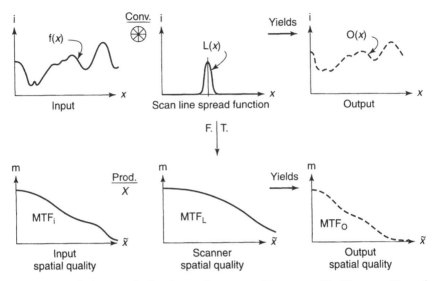

Fig. 2.13 Equivalent methods of assessing scanned image quality: Top row: Convolution of the input function by the scanned spread function. Bottom row: product of their Fourier transformed MTFs. Observe yielded loss of detail in top row and loss in spatial frequency in bottom row.

using a scanning line spread function $L(x)$. As with the scanning PSF of Figure 2.4, this yields the reduced detail output function $O(x)$. Although this is conceptually enlightening, a fundamental alternative is represented in the lower half of Figure 2.13: the theoretical equivalent of *multiplying* the input spectrum MTF_i by the MTF_L of the line spread function. Although the phrase "modulation transfer function" applies rigorously to the Fourier transform of the point or line spread functions, it has been extended for system analysis to include cascaded stages, such as the input MTF_i and the MTFs due to such factors as aberration, image motion smear, aerodynamic perturbation, and, of course, the final output system MTF_o. Exercising this concept entails application of the principle of MTF multiplication. To be valid, this must be limited to those cases where the MTFs are independent, uncorrelated, and incoherent with each other. In the general case, one must include the PTF.

In this regard, it is important to note that in the field of laser scanning, where the light source may be effectively coherent, this analysis employs the incoherent MTF of the scanning spot. That is, we Fourier transform the *intensity* point (or line) spread function; not the amplitude function. Although the scanning spot may, in fact, be formed of coherent light, its sampling of adjacent elements at even infinitesimally

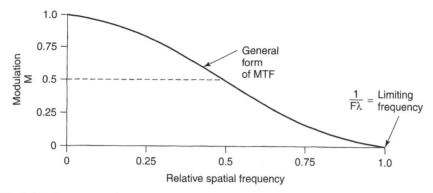

Fig. 2.14   Representative modulation transfer function (MTF). Actual function may vary, while maintaining a monotonic decrement (not maintained, e.g., in certain cases of defocus and aperture obscuration). The spatial frequency at $M = 0.5$ can denote a single point descriptor of MTF quality. Limiting spatial frequency = 1/Fλ, where F = f-number and λ = wavelength.

different times destroys its coherent relationship to its neighbors. Although there are instances when at least a partially coherent MTF need be assumed in laser scanning, these are limited to those cases where a group of coherent spots scans an information medium simultaneously, as in the Scophony process (Section 4.8.3.1).

A representative MTF with typical notation appears as Figure 2.14. Although the actual function may differ from the illustrated Gaussian-like form, as in Figure 2.11, it maintains a generally monotonic decrement. The dashed line at $M = 0.5$ denotes the practical choice for a single-point descriptor of MTF quality determining thereby the spatial frequency at which modulation equals 0.5. (See Section 3.2 and Fig. 3.2.) The spatial frequency limit of 1/Fλ, where the MTF goes to zero, relates to the scanning spot size. The smallest detectable cyclic signal occurs when the scanning spot subtends one spatial cycle; canceling its alternating components. With the spot width given by $\delta = a$Fλ of Equation 2-25, this occurs at the spatial frequency of 1/Fλ.

CHAPTER 3

SCANNED RESOLUTION

3.1 INFLUENCE AND SIGNIFICANCE OF SCANNED RESOLUTION

In the earlier chapters, a few key items were introduced that were to be expanded here. They include the resolution invariant of Section 1.4, the aperture shape factor of Section 2.1.2, and the Gaussian spot size of Section 2.3.2. This chapter develops scanned resolution for general application. The exposition of this subject before concentrating on the scanning techniques underscores its basic independence from any specific methods of deflection or system architecture.* Furthermore, it may be appreciated that the resolution of a scanning system can be calculated straightforwardly, avoiding diversion by potentially complicating system factors.

3.1.1 Basis of Scanned Resolution

Various technical communities apply the word "resolution" differently. In astronomy, it means the limiting small angle that subtends two observed remote objects (in fractional arc seconds or microradians).[‡]

---

* An extensive discussion of scanned resolution appears in Section 2.8 of *Holographic Scanning* (Beiser, 1988), deriving and illustrating several aspects of the topic presented here.
‡ Similarly for pointing accuracy or small angle "staring" (Section 4.5.4), where the descriptor "finesse" is appropriate.

In photographic imaging, it means the largest number of spatial cycles countable in one millimeter of the information medium (in cycles or line pairs per millimeter). Television technology departed from the above "limited field" descriptors by accounting for the *effort* expended (in electronic image capture and display and in signal bandwidth), by defining resolution as the *total number* of picture elements (pixels or lines) conveyed in one scan direction of a total field. Television also counts two pixels per information cycle. This general philosophy is sustained in optical scanning, representing the achievement of full field resolutions that can extend into the tens of thousands of information elements per scan. Certain scanning systems (to be described) can convey 100,000 or more elements per single scan. Usually, the scanned spatial path is effectively linear and traversed at relatively uniform velocity, encompassing data elements that are nominally equally spaced. Scanned resolution is represented by the letter N.

Two fundamental forms of optical scanning, *translational* and *angular*, can exist independently or jointly. Exemplifying these options with a laser beam, translational scan displaces a focal point of spot size δ over a format length S (as by movement along a collimated beam axis of a "pick-off" mirror coupled to a focusing lens). Assuming that elemental spacings correspond to the spot size δ, the resolution of *translational* scan N_s is expressed by

$$N_s = \frac{S}{\delta} \qquad\qquad (3\text{-}1)$$

Angular scan is achieved by nutating a beam that is ultimately converged to focus, such that the locus of its focal spots δ traverses a format width W. Again assuming that elemental spacings correspond to the spot size δ, the resolution of *angular* scan N_Θ may be represented similarly by

$$N_\Theta = \frac{W}{\delta} \qquad\qquad (3\text{-}2)$$

An equivalent angular relationship, introduced in Section 1.4, is given as the ratio of the full active scanned angle Θ to the diffracted angle $\Delta\Theta$. This is analogous to the basic expression for (electrical) signal-to-noise ratio; the ratio of full signal S to the uncertainty in signal ΔS. Equation 3-2 thus extends to

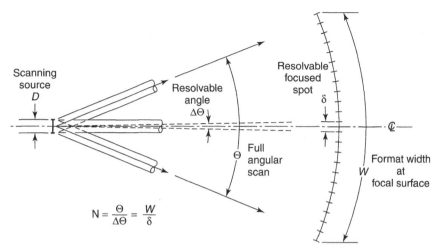

Fig. 3.1 Development of angularly scanned resolution. Scanning source of width D generates full angular scan Θ, which is typically much greater than the resolvable angle $\Delta\Theta$. Scanned Resolution is $N = \Theta/\Delta\Theta$, corresponding to W/δ, the ratio of the format width to the focused spot size.

$$N_\Theta = \frac{W}{\delta} = \frac{\Theta}{\Delta\Theta} \tag{3-3}$$

Figure 3.1 illustrates a source of width D generating a full scanned angle Θ that is typically much greater than the small diffracted angle $\Delta\Theta$ emanating from the aperture D. $\Delta\Theta$ represents the resolvable angle, appearing above in Equation 2-16 and expressed in this notation as

$$\sin \Delta\Theta \simeq \Delta\Theta = a\frac{\lambda}{D} \tag{3-4}$$

The aperture shape factor a, introduced in Section 2.1.2, is discussed further in Section 3.2. Adapting the exemplary case of a converging Gaussian beam (per Fig. 2.9), this diffracted component $\Delta\Theta$ emanating from aperture D propagates over the focal distance f to form the finite spot size δ at the focal plane. With $D \gg \lambda$, then $\sin\Delta\Theta \simeq \Delta\Theta$, and $\delta = f \cdot \Delta\Theta$, validating the denominators of Equation 3-3.

The numerators correspond similarly when forming a flat image of width W. With the use of a flat-field $f \cdot \Theta$ "scan lens," where effectively $W = f \cdot \Theta$, one transforms an arcuate scan to a flat surface. These lens design factors are discussed in Section 4.4.

Substitution of Equation 3-4 into Equation 3-3 provides the most revealing and most useful representation of angularly scanned resolution,

$$N_\Theta = \frac{\Theta D}{a\lambda}$$

(3-5)

The numerator terms Θ and D are the principal variables, whereas the denominator $a\lambda$ often forms a system constant. The importance of Equation 3-5 is manifest in the subsequent discussion that includes the development of the resolution invariant and of resolution augmentation. Angular scanning is used more frequently than translational scanning, for it can provide the highest combination of speed and resolution. It avoids the inertial limitations of reciprocating mechanisms (see Section 4.2) that are often integral features of translational scan.

The most significant aspect of Equation 3-5 is identification of the two principal variables Θ and D, *established at the deflector*, whose product determines the resolution N. Equation 3-5 includes *no explicit term for the spot size* and *invokes no relationship to the imaging optics* following the scanner. Representing diffraction-limited performance in centered paraxial (sin $\Theta \simeq \Theta$) linear optical systems [Lev], Equation 3-5 is independent of the spot size and all other parameters such as format width, relay optics, magnification, and lens focal lengths. Thus angularly scanned resolution can be determined accurately with no need for diversion and possible error by analyses of subsequent optical systems.* Once the diffraction-limited performance is so determined, systematic aberrations and other operational parameters such as scan linearity need be considered and controlled, as is conducted for any practical design.

3.1.2 Resolution Nomograph

Equation 3-5 is plotted in Figure 3.2 as a nomograph, for convenient estimation of scanned resolution. The beam deflection angle Θ appears directly in real degrees. The other main variable, the aperture width D,

---

* Equation 3-5 applies as well to phased array beam steering (Section 4.10.3), in which that Θ may not be assumed an independent variable. Substituting for Θ that due to array diffraction yields a correct Equation 4-51 for N that is not recognizable as being derived from Equation 3-5.

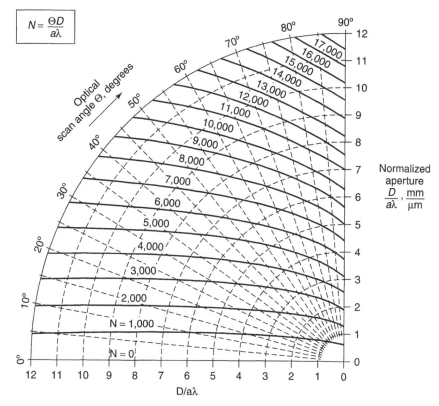

Fig. 3.2 Nomograph of angularly scanned resolution. The optical scan angle Θ is in real degrees, and the radius plots normalized aperture width, $D/a\lambda$. When $a\lambda \simeq 1\,\mu m$, the radius reads D directly in millimeters. Resolution N (bold lines) reads at intersection of scan angle Θ and aperture width $D/a\lambda$. For example, when $\Theta = 50°$ and $D = 8\,mm$ (for $a\lambda \simeq 1\,\mu m$), $N \simeq 7000$ elements per scan.

is plotted in the radial direction. Its notation $D/a\lambda$ includes the essentially constant system denominator of Equation 3-5. The values of resolution N (at the bold crossing lines) are scaled for D in millimeters and $a\lambda$ in micrometers. When $a\lambda$ may be approximated at $1\,\mu m$, (typically in the visible range), the value of D reads directly in millimeters. For example, given $\Theta = 50°$ and $D = 8\,mm$ (for $a\lambda = 1\,\mu$), then $N \simeq 7000$. Also, multiples of either scale yield corresponding multiples of resolution N. Thus miniaturizing the above example, when $\Theta = 5°$ and $D = 0.8\,mm$, $N \simeq 70$. As noted above, these diffraction-limited resolutions merit consideration of and correction for systematic losses due to, for example, aberration.

3.2   APERTURE SHAPE FACTOR

With identification of Equation 3-5 as the resolution equation, its effective deployment merits clarification of the aperture shape factor in the denominator.

This factor, a, was introduced near the end of Section 2.1.2 for an aperture width D illuminated by a Gaussian beam. The discussion in Section 2.1 relates to the differing point spread functions (focused spot contours) resulting from illuminating different aperture shapes. Although the principal lobes of these different spots exhibit a generically similar contour, their absolute widths are different. These differing "spot sizes" are accommodated in the resolution equation by the aperture shape factor. The values of a are determined for particular aperture illumination conditions and spot measurement criteria. Typically, the value of a ranges as $0.75 < a < 2$. Functionally, it represents the amount by which the aperture width D must be widened to render an MTF (at midrange spatial frequencies) equivalent to that for the uniformly illuminated rectangular aperture. Values of a for typical uniformly illuminated and Gaussian-illuminated apertures are summarized and tabulated in Section 3.2.2.

3.2.1   Uniformly Illuminated Apertures

Referring to Section 2.1, one may now identify several aperture shape factors. In Section 2.1.1 two prominent shapes are discussed that are illuminated with uniform intensity distribution. The first is the rectangular or square aperture, and the second is the round or elliptic aperture. The resulting focused spot contour derived from the rectangular aperture of width D is represented in Figure 2.1, where the first null of the main lobe occurs at $1.0(F\lambda)$. With the observation that all focused spot sizes may be generalized to the relationship $\delta = a(F\lambda)$, in which F and λ represent the converging f-number and the wavelength respectively, the value of a for the uniformly illuminated rectangular aperture is taken as

$$a_\square = 1.0 \tag{3-6}$$

The focused spot of the uniformly illuminated round aperture of width D is functionally very similar to that illustrated in Figure 2.1. It exhibits, however, a first null at $1.22(F\lambda)$. Thus considered 1.22x wider than that of a uniformly illuminated rectangular aperture, the value of a for this case is often taken as

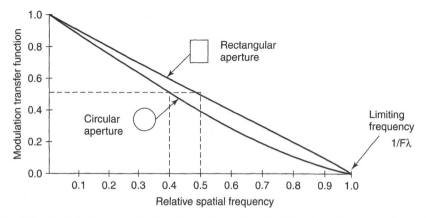

Fig. 3.3 Modulation transfer functions for uniformly illuminated rectangular and circular apertures.

$$a_\circ = 1.22 \qquad\qquad (3\text{-}7a)$$

At the end of the discussion in Section 2.42 of the MTF, it was observed that the relative spatial frequency at the midvalue of MTF = 0.5 represents a good single quality descriptor. Accordingly, Figure 3.3 shows the MTF of the rectangular and circular apertures plotted on the same scales. As expected, that for the rectangular aperture, having a triangular shape per Figure 2.1.2, connects the end points with a straight line. Denoting R as the relative spatial frequency at which MTF = 0.5, then $R_\square = 0.5/F\lambda$ for the uniformly illuminated rectangular aperture. This case is standardized as $a_\square = 1.0$ per Equation 3-6. The MTF curve for the circular aperture shows a reduced midrange because of its increased spot size. The R value for the circular aperture compares to that of the rectangular aperture in the ratio of approximately 0.4/0.5. Thus $R_\circ = R_\square/1.25$, representing an a value of

$$a_\circ = 1.25 \qquad\qquad (3.7b)$$

This signifies an operational loss slightly greater than the 1.22 factor given by Equation 3-7a, which was determined from the first nulls of their point spread functions. This 0.5 MTF criterion for the aperture shape factor [Bei3] is applied below to aperture shapes of less prominent contours.

3.2.1.1 *Keystone and Triangular Apertures*   Among the rotational optical scanners discussed in Chapter 4, there appears the

family of *pyramidal* scanners, whose multifacet surfaces form an apex. As may be perceived, overillumination (overfilling) of such a pyramidal array "head on" (along its rotational axis) results in truncating or apodizing the beam that is reflected from each facet. It forms a triangular boundary or, with an axial obscuration (such as a bearing mount), a keystone-shaped contour. This truncation (of essentially uniform intensity distributions) underwent analysis of aperture shape factors during the mid-1960s [Bei1, Appendix 1] and is summarized below.

As discussed in Section 2.4.2 and represented in Figure 2.12, a powerful method of determining the MTF resulting from scanning an illuminated aperture (or pupil) is to convolve the pupil function $A(\xi)$ with itself (its complex conjugate). This is termed autocorrelation. That is, *the modulation transfer function is the autocorrelation function of the pupil function* [Bro]. For pupil functions that may not be expressed readily in closed form, this is accomplished geometrically by computing and plotting the relative overlapping area of the two apertures (starting at full overlap $\equiv 1$) as a function of the fraction of shift of one aperture (in the scan direction) to the end point of insignificant overlap. This plot is the MTF. A set of four MTFs is illustrated in Figure 3.4. The upper two plots are replots of those from Figure 3.2 of the uniformly illuminated rectangular and circular apertures. With the rectangular aperture serving again as a reference, then as before, $R_O = R_\square/1.25$, yielding $a_O = 1.25$. The keystone aperture (dot-dash) curve results from geometric computation of a shifting overlap area, yielding $R_\triangledown = R_\square/1.5$; making $a_\triangledown = 1.5$. This typical keystone shape is represented by a full width D and narrow width $D/2$. Finally, the triangular aperture (dotted) curve was analyzed by similar (and simpler) computation, yielding $a_\triangledown = 1.7$. All MTFs are calculated in the scanning (and shifting) directions of the width D.

3.2.2   Summary of Aperture Shape Factors

The Gaussian and uniform intensity distributions are the most prevalent illumination functions. Corresponding aperture shape factors are listed in Table 3.1. The Gaussian a values stem mainly from Section 2.1.2, whereas data for the uniformly illuminated conditions derives from this section. Scanning is in the direction of width D. One-dimensional Gaussian distributions (i.e., Gaussian in one axis and constant in quadrature) and their corresponding a values are most applicable to acoustooptic [Bei2] and phased array deflection, as discussed in Section 4.10.3.

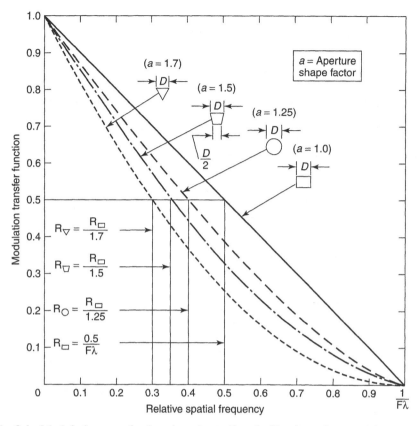

Fig. 3.4 Modulation transfer functions for uniformly illuminated rectangular, round, keystone, and triangular apertures.

TABLE 3.1   Aperture Shape Factor a

Gaussian Illuminated			Uniformly Illuminated		
Spot overlap	a-Untruncated $(D_a \geqq 1.7D)$	a-Truncated $(D_a = D)$	Shape	$\rightarrow |D| \leftarrow$	a
At $1/e^2$ intensity	1.27	1.83	Rectangular	▢	1.0
At $\frac{1}{2}$ intensity	0.75	1.13	Round/elliptic	◯	1.25
For 50% MTF	0.85	1.38	Keystone	▽	1.5
			Triangular	▽	1.7
Beam of width D at $1/e^2$ intensity, centered within aperture boundary of width D_a.			Aperture width D for 50% MTF.		

Data for the Gaussian-illuminated aperture appears in two forms, representing, two illumination conditions: "untruncated" and "truncated." As introduced in Section 2.1.2, to derive maximum resolution from the Gaussian beam, it is necessary that the scanning aperture be large enough to accept the (essentially) full Gaussian function, to include its skirts to where their intensity drops to under 1% of maximum. As shown in Figure 2.2, the beam is considered untruncated when the limiting aperture width $D_a \geq 1.7 D$, where D is the diameter of the Gaussian beam at its $1/e^2$ intensity points. A popular alternate condition that trades off resolution (and throughput optical efficiency) is operation in the truncated condition, where the scanning aperture width is reduced to $D_a = D$, equal to the Gaussian beam width at its $1/e^2$ intensity values. Observing from Table 3.1 the ratio of their a values when the Gaussian beam is so truncated, compared to untruncated, the resolution is degraded typically by a factor of about 1.5x.

3.3   THE RESOLUTION EQUATION, THE RESOLUTION INVARIANT, AND BEAM PROPAGATION

The concluding equation 3-5 of Section 3.1 expresses the basic form of the resolution equation, represented as $N_\theta = \Theta D / a\lambda$ elements per scan. As introduced in Section 1.4, the denominator terms a and λ are typically system constants, where the numerator terms Θ and D are the principal variables, namely, *the total angle Θ through which a beam of width D is scanned*, yielding Equation 1-3. This leads to the resolution invariant of Equation 1-4.

$$I_N = \Theta D = \Theta' D' \qquad (3\text{-}8)$$

This fundamental relationship is exemplified in Figure 1.7, where the deflection of an input beam of width D through an angle of $\pm\Theta/2$ is illustrated traversing a pair of (beam compressor) lenses. The resulting width D' of the compressed output beam is reduced by the same factor as its deflection angle $\Theta'/2$ is increased, maintaining their product constant. Because the ΘD product is proportional to scanned resolution N, Equation 3-8 represents this *conservation of angularly scanned resolution throughout optical transfer.*

3.3.1   Propagation of Noise and Error Components

Not only does the resolution invariant impact the scanned resolution, shown established *at the deflector* and sustained throughout subsequent optical transfer, but the principle applies as well to such factors as the propagation through the optical system of noise and other superposed error components.

Consider, for example, rotary bearing noise as an error component. Let the bearing nonuniformities impart small angular perturbations to a reflective deflector, such as a vibrating mirror or polygon facet that is intended to render x-scan only. Such errors may be resolved in two directions; along scan and cross scan. Although the useful x-scan component Θ_x appears properly in the along-scan direction, the perturbing angular error component ϑ_x appears there, too. Because this deflector is to serve the x-dimension only, it renders no cross-scan (y) information. That is, Θ_y (of x-scan) $= 0$. However, its cross-scan "wobble" angular error ϑ_y (typically of similar magnitude as ϑ_x) remains finite.

A criterion for spot position error is the deviation in the nominal spacing of elemental spots, that is, the fractional misplacement of one of the N elements per scan. Because $\Delta\Theta_x$ of Equation 3-4 represents the nominal angular beam spread (and is assumed equal to the spot spacing), then for reasonable spot position integrity, the error angle ϑ_x is required to be less than $\Delta\Theta_x$. (More critical system criteria may require that $\vartheta_x \ll \Delta\Theta_x$). In the cross-scan direction, raster-forming scan lines are to be disposed spaced equally by an independent mechanism. Because x and y information elements are typically spaced equally, this requires that the cross-scan error ϑ_y must also be less than $\Delta\Theta_x$ (more critically, $\vartheta_y \ll \Delta\Theta_x$), as governed by the same accuracy criterion as for the along-scan direction.

In the x-direction, the desired spot scan angle Θ_x and the perturbing spot scan error angle ϑ_x are inseparable. That is, following the resolution invariant of Equation 3-8, the desired resolution component $I_{Nx} = \Theta_x D_x = \Theta_x' D_x'$ is conserved and the undesired error (ε) component $I_{\varepsilon x} = \vartheta_x D_x = \vartheta_x' D_x'$ is also conserved. Angular error, which is a constituent of the desired scan component, propagates in the same manner as the desired component. In the y-direction, however (in this example), the x-deflector provides no y-informational scan. Because $\Theta_y \equiv 0$, the incidental y-error component ϑ_y may be separated and controlled independently. This is discussed after a related condition, below.

When an angular error results from a perturbation that *precedes* the scanner, there is an opportunity for independent control of the error. This may be accomplished by placing the error-generating component into a smaller beam region, which requires subsequent expansion to fill the deflector aperture properly. This condition is discussed in Section 4.3.5.5 for the case of image derotation. In this way, following Equation 3-8, its angular error ϑ is reduced to ϑ' by a factor equal to the beam expansion ratio by which D is enlarged to D' to attain the required beam size.

Returning to the case of the rotary bearing noise whose cross-scan error component ϑ_y is finite while the information component $\Theta_y = 0$, one can simulate the process described above by *reducing* the beam height D_y anamorphically at the deflector (compared to an original isotropic beam diameter D_x), and then, in a manner similar to the above procedure, reexpanding the beam height beyond the deflector to restore beam isotropy. Subsequent focusing will yield a focal spot whose position is stabilized cross-scan by a factor equal to the initial beam compression ratio. This important process is discussed further in Section 5.1.2.1 "Anamorphic Error Control." Its understanding allows for instituting an amount of correction that is appropriate to the task, avoiding excessive correction that may require the control of resulting introduced optical aberrations.

3.4   AUGMENTED RESOLUTION

The above development of angularly scanned resolution yielded the basic relation represented by Equation 3-5 and the nomograph of Figure 3.2. Equation 3-5 is rigorous for the case of the deflecting beam of width D positioned at the fulcrum of its total angular change Θ, as for the angular scanners illustrated in Chapter 1. Or it is equally accurate when deflecting a *collimated* beam into a flat-field lens. However, it requires adjustment for, for example, a mirrored polygon or a holographic scanner deflecting a beam that converges toward focus. The distinction is that in the latter cases, the deflecting component (e.g., polygon mirror) is displaced from the rotating axis of the scanner while the deflected beam is converging or diverging to or from focus (not collimated). Equation 3-5 is augmented [Bei4] to yield more general relationships that were expanded further as applied to holographic deflection [Bei1].

Consider Figure 3.5 illustrating a deflecting aperture D positioned a distance r from its fulcrum, which rotates about point o.* This is representative of a polygon deflector of radius r (to the facet) rotating about point o through angle Φ such that the beam width D converges

---

* Although Figure 3.5 appears to illustrate the output beam in a plane that is normal to the rotating axis, it represents more generally the output beam *as projected* on a plane that is normal to the axis. This allows accounting for the resolution of a system whose output beam departs from the normal plane.

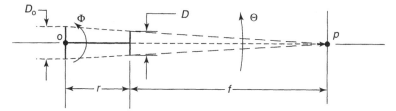

Fig. 3.5 Basis of augmented resolution. Deflecting source of width D, displaced from axis o by radius r, propagates a converging beam over focal distance f to point p. During scan, effective larger aperture D_o renders resolution augmentation. Correspondingly, movement of point p during scan is augmented by transverse movement of real source D.

to a point p over distance f while scanning through angle Θ. Point p not only is displaced because of the scan *angle* Θ but also displaced because of the *translation* of aperture D over the distance $r\Phi$. The resolution N is now determined by the sum of the two basic processes represented by Equations 3-1 and 3-2, expressed by

$$N = N_\Theta + N_s \qquad (3\text{-}9)$$

An heuristic perception of resolution augmentation appears directly when considering in Figure 3.5 that the output beam derives effectively from a *larger* aperture D_o located at the fulcrum o. By similar triangles, $D_o = D(1 + r/f)$, which when treated as D in Equation 3-5 yields,

$$N = \frac{\Theta D}{a\lambda}\left(1 + \frac{r}{f}\right) \qquad (3\text{-}10)$$

This corresponds to Equation 3-9 as the aperture D is effectively translated through transverse distance $S = r\Phi = r\Theta$ ($\Theta = \Phi$ in this case). This adds the augmenting component N_s represented by Equation 3-1. (The right-hand term of expanded Equation 3-10 is $N_s = \Theta Dr/af\lambda$, which reduce to Equation 3-1 via Equation 2-17). The case of $\Theta = \Phi$ represents the condition identified as "radial symmetry," discussed below. Equation 3-10 is then adapted further to include the condition of $\Theta \neq \Phi$.

3.4.1 Radial Symmetry and Scan Magnification

Certain rotational scanner architectures exhibit a basic characteristic identified as radial symmetry. This is determined by the relationship between the incident illuminating beam (or passive radiant flux) and the rotational axis of the scanner. *When the illuminating beam converges to or diverges from the rotating axis of an angular scanner, the system exhibits radial symmetry* [Bei1]. Figure 1.10 illustrates this condition with an active illuminating beam convergent upon the monogon mirror and focused on the axis at point o. Also, when the incident beam is collimated and parallel to the rotating axis of a similar scanner (Fig. 1.9), this forms a special case of radial symmetry, for the beam derives effectively from a very distant point on the axis. Figure 3.6 illustrates a pyramidal polygon operating in this manner. Scanners that operate in radial symmetry deflect the beam through an angle Θ that is equal to the mechanical scan angle Φ. That is, defining m as the ratio of the optical-to-mechanical angular changes, radial symmetry occurs when $m = d\Theta/d\Phi = 1$. The parameter m is called the *scan magnification*, ranging in value between 1 and 2, depending on scanner type and its illumination. At uniform angular velocities, $m = \Theta/\Phi$.

The prismatic polygon in a typical preobjective architecture, illustrated in Figure 3.7, exhibits $m = 2$ when the illuminating beam resides in the same x-z plane as the scanned beam, normal to the rotating axis. This is an extreme departure from radial symmetry, as is its single-mirror prototype of Figure 1.8. When the illuminating beam is

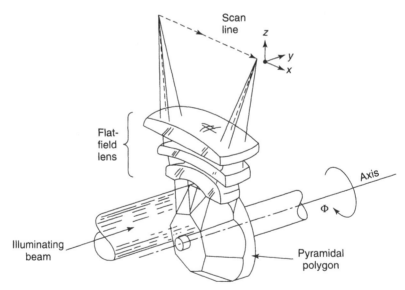

Fig. 3.6 Pyramidal polygon operating in radial symmetry. Input beam, parallel to the axis, derives effectively from a very distant point (at $-\infty$) on the axis. In radial symmetry, the beam reflected from the facet scans an angle Θ, which is equal in magnitude to the polygon rotational angle Φ.

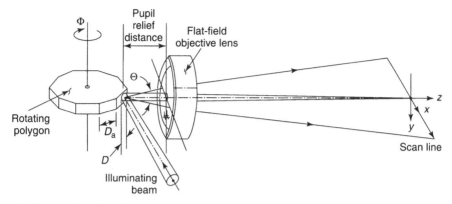

Fig. 3.7 Prismatic polygon operating in extreme radial asymmetry. Input beam illuminates polygon in plane normal to the rotating axis. For facets parallel to the rotating axis as shown, the scanned angle Θ also resides in the plane normal to the rotating axis and its magnitude is double the rotational angle Φ.

converging or diverging, the parenthetic augmenting term of Equation 3-10 requires adaptation to the value of m. Following parametric analysis, it is shown [Bei1, Bei4] that with the introduction of m into Equation 3-10

$$N = \frac{\Theta D}{a\lambda}\left(1 + \frac{r}{mf}\right) \quad \text{elements per scan} \quad (3\text{-}11)$$

it provides an accurate measure of augmented scanned resolution (within 1% over a very wide range of operating conditions.) A rigorous form of Equation 3-11, from which it was derived, is obtained by replacing m with m' [Bei4], where

$$m' = \frac{\sin \Theta_{1/2}}{\sin \Phi_{1/2}} \quad (3\text{-}12)$$

in which the subscripts $\frac{1}{2}$ denote the half-angles from the centers of deflection.

A useful variation of Equation 3-11 expresses the magnification m and the augmenting term r/f as more independent modifiers of the resolution N,

$$N = \frac{\Phi D}{a\lambda}\left(m + \frac{r}{f}\right) \quad \text{elements per scan} \quad (3\text{-}13)$$

In applying Equations 3-10, 3-11, or 3-13, the focal distance f (represented in Fig. 3.5) is of positive value for a convergent beam, and infinite for a collimated beam. Thus resolution elements are added to or subtracted from the conventional angular scan value of Equation 3-5, depending on whether the output beam is convergent or divergent, respectively. And, if the beam is collimated, the augmenting term disappears. The effects of radial symmetry and augmented resolution are reviewed in Chapter 4 as they relate to exemplified rotational scan systems.

3.4.2   Augmented Resolution for Holographic Scanners

Although discussion of holographic scanning appears below in this work (Section 4.4), the interpretation of m for this type of scanner merits attention here. As expressed above, many forms of holographic scanning follow very closely the architecture of polygonal scanners, to the extent that those in radial symmetry exhibit $m = 1$. Examples are provided in Section 4.4 that exhibit $r =$ a finite value, emulating the pyramidal polygon. Also, the output beam may be collimated or focusing, forming the augmenting term r/f described above.

A family of holographic scanners appears in transmissive disk form. Typically, the incident and diffracted beams form Bragg angles β_i and β_o with respect to the grating normal, per Figure 3.8. The value of scan magnification m develops [Kra1, Bei1] as a function of these angles and is represented by the grating equation,

$$m = \sin\beta_I + \sin\beta_o = \lambda/d \qquad (3\text{-}14)$$

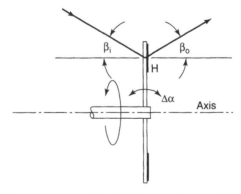

Fig. 3.8 Holographic scanner operating in Bragg regime such that beam angles with respect to the hologram H normal form $\beta_i = \beta_o = \beta$. Then, scan magnification is represented by the grating equation, $m = 2\sin\beta = \lambda/d$, in which d is the grating spacing.

in which d is the grating spacing. For example, when $\beta_i = \beta_o = 45°$, $m = \sqrt{2}$ and when $\beta_i = \beta_o = 30°$, $m = 1$.

3.5 RESOLUTION IN PASSIVE AND REMOTE SENSING SYSTEMS

The basic expression for angularly scanned resolution given by Equation 3-3 is restated here as Equation 3-15:

$$N_\Theta = \frac{\Theta}{\Delta\Theta} \qquad (3\text{-}15)$$

in which Θ is the full scanned angle and $\Delta\Theta$ is the resolvable angle. It applies equally to passive and remote sensing systems, with the adaptation of $\Delta\Theta$ to non-diffraction-limited systems. In remote sensing, the

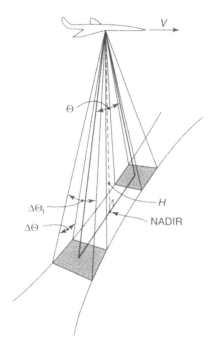

Fig. 3.9 Geometry of an airborne line-scanning system. Aircraft transports the scanned function along "track" at velocity V at height H over the nadir on the scanned surface. Total "field of view" is Θ, compared to the "instantaneous field of view" $\Delta\Theta$, forming scanned resolution $N_\Theta = \Theta/\Delta\Theta$.

smallest detectable angle may be determined by the smallness of the photodetector element, represented in Figure 1.2 for generic passive scanning. When the object space of Figure 1.2 is a very long distance, as in remote sensing, the arriving flux from the object is essentially collimated, locating the detector one focal length from the objective lens. Rather than being diffraction limited per Equation 3-4, now $\Delta\Theta = d/f$, in which d is the detector element width (in the scan direction) and f is the objective lens focal length [B&J]. Furthermore, the arriving illumination may be assumed to be of uniform intensity distribution on the scanner and/or lens aperture.

Figure 3.9 illustrates the geometry of an airborne line-scanning system. The total scanned "field of view" is Θ, and its resolvable (elemental) angle in that direction is $\Delta\Theta$. The aircraft transports the scanned function along "track" at velocity V at the height H over the nadir on the scanned surface. The "instantaneous field of view" is identified as the scanned solid angle $\Delta\Theta \cdot \Delta\Theta_t$, where $\Delta\Theta_t$ is the resolvable angle in the track direction. To assemble contiguous scan lines (accounting for the apparent "bow-tie" distortion) [B&J], the equivalent angular velocity of the scanning system in the track direction is $V/H = \dot{s} \cdot \Delta\Theta_t$, where $\dot{s}$ is the scan rate in scans/s. For a system with n detectors aligned in the track direction, $V/H = n\dot{s} \cdot \Delta\Theta_t$.

CHAPTER 4

SCANNER DEVICES
AND TECHNIQUES

4.1 SCANNING TECHNOLOGY ORGANIZATION

Figure 4.1 classifies optical scanning technology, arranged and extended beyond its original 1974 construction [Bei2]. Although some inevitable expansion that developed over the last three decades is represented and clarified here, the basic organization is sustained.

The most durable concept is the division of the technology into two principal categbries, namely, "high inertia" and "low inertia." The distinction between the two relates primarily to scan function flexibility, that is, between repetitive and agile control. The high-inertia systems maintain high regularity. They resist rapid changes in scan speed and in the locus of the scanned function. The low-inertia systems, however, allow timely control and alteration of a scan trajectory. They often exhibit "random access" capability.

A novel representation in Figure 4.1 illustrates this separation of task with the appearance of "oscillatory" under both the high inertia and low inertia categories. The low-inertia oscillatory descriptor leads to the galvanometric devices, as represented alongside the nonmechanical—Acoustooptic and Electrooptic—devices. (Each of these techniques is described in a later discussion.) The galvanometer scanner continues to be identified as low inertia, for its low moment of inertia* rotor is in a sufficiently damped broadband suspension to support variable

---

* Moment of inertia $\equiv mr^2$; m = effective mass at r = radius of gyration.

63

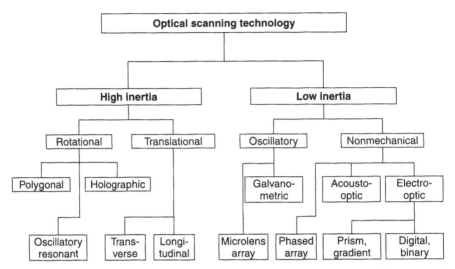

Fig. 4.1 Classification of optical scanning Technology. After "Laser scanning systems," by L. Beiser in *Laser Applications*, Volume 2, by Monte Ross, ©1974, Elsevier Science (USA), reproduced by permission of the publisher.

velocity and/or variable rate scan cycles. This includes the formation of the important "sawtooth" ramp having a relatively short retrace interval and the ability to "random access" any point in an image format within its retrace time limitation.

In sharp contrast to the above, the high-inertia "rotational' block leads—perhaps surprisingly—to the oscillatory resonant scanner, in the same domain as the fully rotational polygonal and holographic scanners. Although the resonant scanner may appear physically similar to the galvanometer, it forms a sinusoidal scan function. This unique oscillating frequency is not readily adjustable, and its low-loss resonant suspension sustains this frequency, disallowing rapid start or stop. Thus, although the rotor of the resonant scanner may actually exhibit a low moment of inertia (as does the galvanometer), it exhibits the scanning properties of a high-inertia device. Significant further distinctions between the galvanometer and resonant scanners are presented in Section 4.5.2.

Also added to the high inertia category are the translational devices, such as mechanical transports of the object, the storage medium, or the optics. In most cases, these fall into the category of objective scanning, described in Section 1.5.1.3. They need not be linear transports. An excellent example is illustrated in Figure 1.11 of the above-referenced section: a drum supporting a storage medium rotating under a focusing objective lens that is also translated longitudinally, parallel to the drum axis. Capable of forming an extremely precise helical scan, this important 'external drum" system articulates in a manner similar to the machinist lathe—and, notably, similar to the Edison cylinder. Further detailing is reserved for subsequent discussion.

4.2 HIGH-INERTIA SCANNING

High-inertia scanning typically entails motion of a significant mass. When properly balanced, such action often provides a bonus by suppressing perturbation of the desired scan function. When operating in continuous rotation, it benefits also from nonoscillatory and nonreciprocating action to render a unique combination of high speed and high uniformity of data point transfer. Rotational components, which include the polygons and holographic scanners, can form highly ordered data points at rates extending into the hundreds of megapixels per second and can be designed to provide pixel spatial uniformities (along scan) that challenge the instrumentation for measurement to such high accuracy.

The linear translational components, such as the traverse mechanisms and position stages, derive position servo feedback typically from linear optical encoders. They can provide extremely high-accuracy linear motion at substantive velocities. Although these speeds are outstripped by the flat-fielded rotational scanners rendering linearized scan, the capacity of the translation stages for moving heavy loads and extremely low f-number objective optics allows traversing extremely small pixels at a high data rate. For example, imaging 1-μm spots at a velocity of 2 m/s represents a data rate of 2 mpixels/s. However, transports that form the useful scan in one direction can be required to retrace at a much higher speed to form a favorable duty cycle (Section 4.2.3). This burdens decelerating and reaccelerating massive moving members, limiting traverse mechanisms to relatively slow scan retrace periods.

4.3 ROTATING POLYGONS

As introduced in Section 3.4.1, rotating polygons are classified in two major configurations: the pyramidal polygon of Figure 3.6 and the more utilized prismatic polygon of Figure 3.7. Although the dominant characteristic of both polygon forms is the regularity of the facet-to-facet (polar) angle and of the facet-to-axis angle, some are designed to be nonuniform in either or both parameters [She]. Nonuniform polar angle separation between facets yields nonuniform temporal and spatial scanned beam lengths (in the "along-scan" direction), whereas nonuniform angular orientation of the facets with respect to the rotating axis yields variations in scan line positioning (in the "cross-scan" direction). These options are rarely implemented intentionally. Most are designed for uniform facet angles in both directions. The prospect of encountering residual (nonintentional) angular deviations is, however, real—resulting from the application of practical machining accuracies and fabrication economies. The consequences of such "manufacturing tolerance" imperfections are the small (but sometimes critical) nonuniformities in scan line timing, length, and placement. Corrective action that may be taken to abate these errors is discussed in Chapter 5. A rarely used design option is the inverted polygon [She] having the facets formed on the *inside* surface of a ring-shaped rotating substrate rather than on the outside circumference of a continuous substrate. These may also be constructed with regular or irregular facet angles.

Significant distinctions appear in the operation of pyramidal and prismatic polygons, discussed below. Either form may have any practical number of facets. When the pyramidal type reduces to one facet, such as illustrated in Figures 1.9 and 1.10, it is commonly called a monogon, whereas in remote sensing, it may be called an oblique or single ax-blade scanner. When comprised of two oblique facets symmetric about the axis, it is identified as a wedge, double ax-blade, or knife-edge scanner [B&J, Wol].

4.3.1   Distinctions Between Pyramidal and Prismatic Polygons

Table 4.1 highlights the principal features and distinctions of typical polygon scanners. For example, because of their differing scan magnifications (item 3), a given mechanical rotation angle Φ results in an optical scan angle Θ of the prismatic polygon that is twice that of the pyramidal polygon. Thus, to scan equal optical angles of equal beam widths D (for equal resolutions N) and to realize equal duty cycle η (Section 4.3.2) at equal scan rates, the prismatic polygon requires twice the number of facets, is almost twice the diameter, and rotates at half the speed of the pyramidal polygon. The precise diameter is a function of the aperture shape factor (Section 3.2) and the changing projection of the beam cross section on the facet during its rotation (item 8 and Section 4.3.5.1). Noteworthy along with the doubling of its scan angle

TABLE 4.1 Features of Typical Polygon Scanners

Item	Description	Pyramidal	Prismatic
1	Input beam direction	Radially symmetric[a] (typically parallel to axis).	Perpendicular to axis[b]
2	Output beam direction	Arbitrary angle (typically perpendicular) to axis.	Perpendicular to axis[b]
3	Scan magnification, $m = d\theta/d\phi$	1 ⎫	2 ⎫
		⎬ Operating per above	⎬ Operating per above
4	Along-scan error magnification	1 ⎭	2 ⎭
5	Maximum scan angle	$2\pi/n$ (n = number of facets)	$4\pi/n$
6	Output beam rotation about its axis[c]	Yes	No
7	Optical aperture shape (overfilled)[d]	Triangular/keystone	Rectangular
8	Along-scan beamwidth on facet (underfilled)	Same width as beam	Widened[e]: $D_m = D/\cos\alpha$
9	Error due to axial shift of polygon[f]	Yes	No
10	Error due to radial shift of polygon[f]	Yes	Yes

[a] See Figs. 1.9, 1.10, and 3.6 and Section 3.4.1.
[b] All beams in same plane perpendicular to rotation axis.
[c] See Section 4.3.5.5. Observable when beam is nonisotropic.
[d] See Table 3.1, uniformly illuminated.
[e] α = angular departure from normal landing.
[f] Image focal point shift in noncollimated light; no error in collimated light.

by the prismatic polygon is the doubling of the along-scan angular error (item 4). Note, too, the susceptibility to shift errors due to axial and radial shifts of the polygon in noncollimated light (items 9 and 10).

4.3.2 Duty Cycle

The active portions of scan cycles are almost always buffered by blanking or retrace intervals, allowing time to reestablish the scanning aperture at the start of the new scan. This interval can include short overscan portions straddling the main format that may be used for synchronization and/or radiometric calibration. The ratio of the active portion to the full scan period is termed the duty cycle η, expressed as

$$\eta = 1 - \tau/T \tag{4-1}$$

in which τ is the blanking time and T is the full scan period. For the underfilled polygon scanners, a practical interpretation of the time loss factor in Equation 4-1 is $\tau/T = D_m/D_a$, yielding,

$$\eta \simeq 1 - D_m/D_a \qquad (4\text{-}2)$$

in which (per Fig. 3.7) D_m is the projected beam width *on the facet* (Section 4.3.5.2) and D_a is the full facet width. Thus this depletion in duty cycle is effectively the fraction of the full scan period that the traversing corner of the facet truncates and distorts the beam, preventing its full utility during that time. This limits the *useful* scan angle Θ to

$$\Theta = \eta\Theta_{max} \qquad (4\text{-}3)$$

where Θ_{max} is the scan angle that could be available over the full period T. A reduced duty cycle imposes increased bandwidth for a given average data rate. It also increases the power required to accomodate a given exposure or detection sensitivity. With Equation 4-3 not limited to a particular coordinate (x or y), evaluation of the required exposure energy (or signal gain for detection) may entail determination of the duty cycle in the quadrature direction as well. Such evaluation is necessary, for example, in the case of repetitive two-dimensional (raster) generation for video recording or projection display.

4.3.3   Over- and Underillumination (Over- and Underfilling) of the Facet

Overilluminated (overfilled) facets were introduced in Section 3.2.1 for discussion of the resolution of uniformly illuminated apertures. In Section 3.2.1.1 (illustrated in Fig. 3.4) appear four examples; the first two (triangular and keystone) represent aperture shapes of pyramidal polygons, the third (round or elliptical) is typical of flat vibrational mirrors, and the fourth (square or rectangular) typical apertures of vibrational devices or of prismatic polygons. Overillumination of polygon facets imparts unique characteristics to the scan process. Along with gleaning the highest possible resolution and speed from a given polygon by utilizing its widest possible aperture, it also allows the largest possible duty cycle of 1.0 (100%) as a limit. This is accomplished (Fig. 3.6) by overilluminating two adjacent facets simultaneously, such that the scan by the second facet is initiated at the instant it is terminated by the first facet. The trade-off is the great loss of light flux due to apodization of a wide beam, utilizing only the central portion of a Gaussian-like beam to approach uniform illumination. Further loss is due to selection of a small fraction (at best, one-half) of that residue at any one time, as the facet traverses the (at least twofold) wider and already apodized beam. The round or elliptical beam is a typical candidate for the underfilled case.

4.3.4   Facet Tracking

A method of avoiding much of the loss of illumination due to overfilling the facets is the technique of facet tracking. It entails the extra step of scanning the illuminating beam (of normal width D) by an auxiliary means, in synchronism with the shift of the facet during polygon rotation. Thus the illumination remains seated centrally on the full facet during scan. The facet continues to reflect this beam through its scanned angle in the conventional manner. Not only is much of the illumination conserved, but this is accomplished with a minimal facet width, allowing use of a polygon diameter that is almost as small at that of the overfilled type having the same number of facets. Forms of this general technique are represented by using acoustooptic tracking of a prismatic polygon [D&J], refractive window rotation with a prismatic polygon [Str], and an early configuration of refractive windows rotating with a pyramidal polygon [Mac].

4.3.5   Design Considerations

A significant architectural distinction between the two types of polygons (not charted in Table 4.1) is the relative freedom of orientation of objective optics with respect to the scanner. Comparing Figures 3.6 and 3.7, one sees that the input and output beams of Figure 3.6 are so separated that the objective lens may be placed arbitrarily close to the pyramidal polygon, whereas Figure 3.7 shows a need for strict observance of the pupil relief distance of the objective lens to avoid its truncation of the input beam. A larger pupil relief distance requires use of a larger-diameter objective lens and its attendant increase in size and cost. And, observing the monogon of Figure 1.10, and extending that to a (postobjective) pyramidal polygon, one can also appreciate the option of orienting the input objective lens arbitrarily close to the polygon. These factors relating to pupil relief distance are taken into account in the following discussion with regard to the popular case of the underfilled prismatic polygon.

4.3.5.1   Architecture of the Prismatic Polygon and Its Flat-Field Lens

The underfilled prismatic polygon operating preobjective with its flat-field lens is illustrated in Figure 3.7, and the critical region relating the two is detailed in Figure 4.2. This prominent configuration merits special attention for determining an imposing array of its operating parameters, such as its number of facets, diameter, orientation with respect to the optical axis, and pupil relief distance for specific input beam angle, scan angles, scan duty cycle, and size of the flat-field lens housing. These apparently unrelated major design factors can be determined to high accuracy by utilizing the design procedures represented here and in Section 4.3.5.2.

Following Figure 4.2, the distance P (pupil relief distance) is measured from the scanner facet surface to the lens input surface along the principal axis of the lens. It is determined primarily by the need for the input beam to clear the edge of the lens by distance Ds ($D \equiv$ beam diameter; $s \equiv$ safety factor ≈ 2). A shallower input beam angle (reduced β) increases P (for the same freedom from interferance) and imposes upon the lens a larger aperture—and an increase in its size and cost. However, a wider input beam angle (increased β) forces a larger facet width W because of increased off-axis landing of the beam on the facet (later determined) forming a corresponding increase in polygon diameter. Another trade-off is the scan angle Θ. Although a wider Θ also increases lens complexity for correction of aberration and flat-field operation, it relieves some demand on the precision of the scanner by allowing a greater angular (wobble) error for a given number N of resolution elements within the enlarged Θ scan. These factors must be considered for a specific design objective.

Establishment of the scanner-lens relationships requires an estimate of the polygon facet count, in light of its diameter and speed. Its speed is determined primarily by the desired bandwidth and data rates and then by other operational factors, such as inertial facet deformation, windage, and multiplexing. Diffraction-limited relationships are employed here throughout, requiring practical adjustment for anticipated aberrations in real systems.

Performance objectives that are usually predisposed for a given system are the resolution N, the full optical scan angle Θ (which relates to the format width W), and the duty cycle η. The relationships between N and Θ are the subject of Chapter 3, and duty cycle considerations appear in Section 4.3.2. The values of N and Θ must be determined as practical for the flat-field lens. Typical high values are $N = 20,000$ and $\Theta = 1$ radian, achieving reasonably flat and linear scans over format widths greater than 100 mm. Narrower formats generally sacrifice N

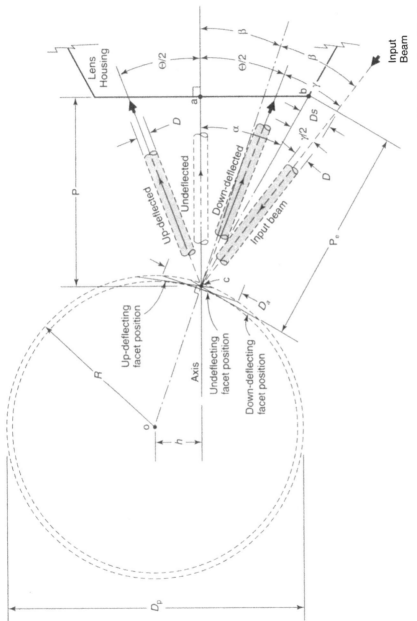

Fig. 4.2 Relationship of prismatic polygon, its input and scanning beams, and its objective lens. Undeflected and limit beam conditions are shown. From L. Beiser, "Design equations for a polygon laser scanner," in *Beam Deflection and Scanning Techniques, Proc SPIE*, Vol. 1454 (1991). Reproduced by permission of the publisher.

and Θ because of their shortened focal lengths, striving for correspondingly lower beam f-numbers to attain the smaller spots. Following these preliminary judgments, from Equation 3-5, the beam width at the deflector is determined as

$$D = Na\lambda/\Theta \qquad (4\text{-}4)$$

in which a is the aperture shape factor. The number of facets n is determined from Table 4.1 and Equation 4-3 as

$$n = 4\pi\eta/\Theta \qquad (4\text{-}5)$$

whereupon it is adjusted to an integer. This establishes the scan efficiency η for a given Θ. Because the number of facets n occurs only in integral steps, optimizing the efficiency η may warrant a minor adjustment of Θ.

4.3.5.2 *Relationships Between Scanner and Lens* The scanner size and relationship to the flat-field lens may now be determined. Figure 4.2 illustrates a typical prismatic polygon operating with all beams in the plane of the polygon substrate (the plane of the paper). One of n facets of width D_a is shown (solid lines) in three positions: undeflecting (neutral) and its two limit positions. The reflected beams are shown in corresponding positions. A lens housing edge denotes the input surface of a flat-field lens. Angle γ provides clear separation of the input beam and the down-deflected beam or lens housing. The pupil relief distance P (perpendicular to the lens housing at axis point a) and its slant distance P_e establish angle α as

$$\alpha = \cos^{-1}(P/P_e) \qquad (4\text{-}6)$$

Angle α is the off-axis illumination on the polygon that broadens the input beam width on the facet from D to D_m. (Note that α approximates the angle between the input beam and the normal to the up-deflecting facet.) An additional safety factor t ($1 \leq t \leq 1.4$) limits one-sided truncation of the beam by the edge of the facet at the end of scan, yielding

$$D_m = Dt/\cos\alpha \qquad (4\text{-}7)$$

Replacing D in Equation 4-2 with D_m and solving for the full facet width D_a,

$$D_a = D_m/(1-\eta) \qquad\qquad (4\text{-}8)$$

which allows for maximum off-axis landing of the input beam and also avoids truncation of the input beam during the end of the active portion of scan.

As developed earlier [Bei5], the polygon outer (circumscribed) diameter $D_P = D_a/\sin(\pi/n)$ may now be expressed with operational parameters of Figure 4.2,

$$D_p = \frac{Dt}{\sin(\pi/n)\cos\alpha(1-\eta)} \qquad\qquad (4\text{-}9)$$

Solution of Equation 4-9 or one of similar form [Kes] entails determination of α, the angle of off-axis illumination on the facet, usually requiring a detailed layout similar to that of Figure 4.2. Aiding system perception, Equation 4-6 and Figure 4.2 reveal that

$$\cos\alpha = P/P_e \qquad\qquad (4\text{-}10)$$

Substituting this into Equation 4-9 yields,

$$D_p = \frac{Dt}{\sin(\pi/n)(1-\eta)} \cdot \frac{P_e}{P} \qquad\qquad (4\text{-}11)$$

still requiring a determinatin of P_e for a given pupil relief distance P. After some accurate small-angle approximations and series expansion of $\cos\alpha$, the relationship for D_p may be expressed [Bei5] with P_e implicit in terms of the principal design parameters,

$$D_p = \frac{Dt}{\sin(\pi/n)(1-\eta)} \cdot \frac{1+\Theta D/P}{1-\Theta^2/8} \qquad\qquad (4\text{-}12)$$

The numerator of the second term, represented originally by $1 + \Theta Ds/2P$ [Bei5], was simplified here by letting the beam spacing factor s take a practical value of 2 for good beam clearance. One may observe also in the numerator of the second term the reappearance of the resolution invariant ΘD, discussed particularly in Section 3.3. The terms in Equation 4-12 are sufficiently basic as initial design parameters to allow reasonable estimate of the polygon diameter. Evaluation of the rigorous Equation 4-9 vs. this equation in typical operational conditions reveals a reduced value by Equation 4-12 of approximately 1.3%.

Proper orientation of the polygon also requires the height h (Fig. 4.2), the distance from the polygon rotational axis o normal to the lens axis. Assume first a typical polygon of relatively high facet count ($n \geq 12$) to which the input beam is directed to the center of the undeflected facet, encountering the inscribed circle of radius R' where $R' = R \cos(\pi/n)$. Because $\sin\beta = h/R'$, then

$$h = R \cos(\pi/n) \sin\beta \qquad (4\text{-}13)$$

However, the intersection point of the input beam at the polygon facet moves during scan as the facet rotates from this center position. As the facet shifts under the beam, the intersection point moves toward the source of the beam, "lifting up" on the corners of the facet [Kle, L&K, Mar1]. Also, because the polygon is illuminated asymmetrically by the input beam (which is displaced from the facet normal by angle β), this intersection of the beam with the facet also becomes more asymmetric at lower facet count (where the arc of the superscribed circle may no longer be assumed equal in length to its chord, the length of the facet.) The consequence of this is analyzed in a recent publication [Mar3].

4.3.5.3 *Overfilling, Double-Pass, and Facet Tracking*   Although discussion in the preceding sections addresses the most popular form of polygon operation, other techniques may be utilized advantageously. An important variation to the above underfilled prismatic polygon is the overfilled polygon (Section 4.3.3), a polygonal adaptation of Figure 1.4, represented in Figure 4.3 in double-pass form. Such operation could be accomplished by illuminating the polygon asymmetrically as in Figure 4.2. However, the symmetry of illumination on the facets (in the scan direction) is advantageous for Figure 4.3; it reduces the angle α (Fig. 4.2) to zero. This option is extremely conservant in polygon diameter, for the full facet serves as the beam aperture. When overilluminated (overfilled), the duty cycle can approach 100% by illuminating two adjacent facets. In contrast, when underfilled, obtaining a high duty cycle requires that the facet be much wider than the beam, incurring a substantial increase in the polygon diameter. Thus Figure 4.3 represents a class of polygon scanners capable of providing the combination of very high resolution and speed simultaneously [Bei2]. Its principal trade-off is the loss of illuminating flux when an input Gaussian beam is expanded sufficiently to overfill two facets while approaching uniform flux density on each active facet. Another precaution is the need for minimizing the angular separation (β; Fig. 1.3) between the

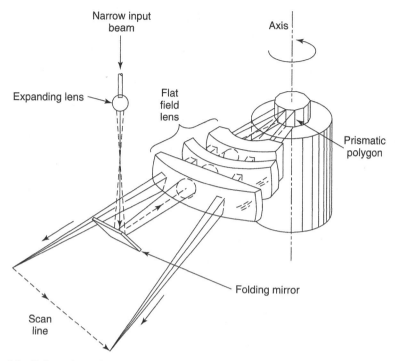

Fig. 4.3 Prismatic polygon in double-pass configuration. Input beam, expanding beyond focus, is directed by folding mirror through flat-field lens, filling polygon facets with collimated light. Scanned reflected beam is refocused by flat-field lens, forming scan line. Input and output beams are skewed slightly (vertically) for separation by folding mirror. From [Bei3].

input and output beams, to allow formation of a usefully straight (unbowed) scanned locus.

The illumination of two adjacent facets by one expanded beam is the most accessible form of filling the facet while attaining a high duty cycle. However, a quest for this advantage without serious loss of light flux is represented by the technique of "facet tracking." This 'simple' principle of scanning the illuminating beam synchronously to track the center of the moving facet has, in practice, required novel beam manipulation technique that impose added complexity. Three methods of facet tracking are identified in Section 4.3.4.

Because overfilling and facet tracking can be applied to pyramidal as well as prismatic polygons, it is appropriate to review the characteristics of the pyramidal polygon, particularly when operating in radial symmetry. [Although it is possible to configure the prismatic polygon in radial symmetry (per Fig. 4.10), it is rarely operated in that mode.]

Referring to Section 3.4.1 and its Figure 3.6, note that in radial symmetry, scan magnification $m = 1$. Also notable is the relative freedom of separating the input and output beams of the pyramidal polygon (see Section 4.3.5), allowing for a minimal pupil relief distance and a correspondingly reduced size of the flat-field lens (in preobjective scan) achieving otherwise similar performance. Also provided is the nominally perfect straight scan line, suffering no bow in propagating axially, through the flat-field optics (in contrast to the possible bow in the above-described double-pass mode of Fig. 4.3.)

4.3.5.4 *Postobjective Configurations*   Because postobjective operation (Section 1.5.1.2) yields typically an arced scan (see Fig. 1.10 vs. Fig. 1.9 and limaçon scan [Bei2]), such curved fields serve well where advantageous. A popular example is the internal drum scanner utilizing variations of the monogon scanner of Figure 1.10. The formation of short linearized scans in postobjective polygon operation is discussed at the close of this section.

A basic multifacet scanner that merits attention is the postobjective pyramidal configuration, illustrated in Figure 4.4. It represents the extension of a rotating monogon to a regular pyramidal polygon having similar illumination. Because the initial focal point intersects the rotating axis, it retains the properties of radial symmetry (Section 3.4.1), directing the scanning beams into focal point loci on a concentric cylindrical surface. For an active input device (i.e., *reading* stored data on the cylindrical surface), one may consider dividing the storage medium into n sectors or strips (where n corresponds to the number of facets.) If each sector is provided with localized light collection and detection, the data on each sector may be read individually—as successive beams of the "rotating crown" sample each sector.* This configuration would render efficient utilization of illumination.

If, however, the system is intended for *writing* on storage media, it is challenging to consider methods of modulating the beam(s) for more than one sector. When writing only one sector, the optical efficiency is depleted on $\eta < 1/n$, that is, wasting all the illumination but that on one facet. One may configure modulating simultaneously two opposite beams, forming two diametric channels, and perhaps even four channels in quadrant sectors, depending on the number of facets. In these "thought experiments," it is reasonable also to consider the options of over- or underfilling of the facets, a discussed above.

---

* Note that this idealized configuration may only approach the condition of n sectors equaling the number n of facets, to allow mounting space for the traversing scanning system to be supported mechanically within the concentric cylindrical enclosure.

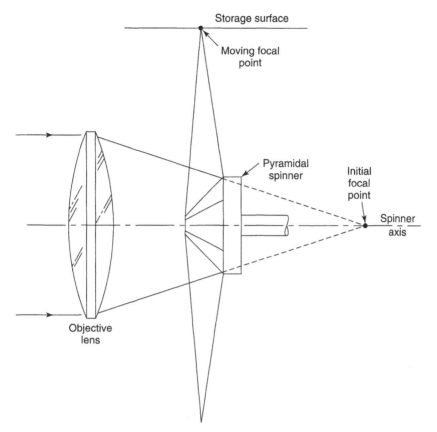

Fig. 4.4 Postobjective pyramidal scanner. Provides radial symmetry when initial focal point intersects spinner axis, forming circular scan locus. From "Laser scanning systems," by L. Beiser in *Laser Applications*, Volume 2, by Monte Ross, ©1974, Elsevier Science (USA), reproduced by permission of the publisher.

In the form illustrated by Figure 4.4, the condition of radial symmetry (which scans a perfectly circular arc) is achieved by converging the initial focal point on to the rotating axis and apparently also by maintaining the axis of the objective lens coaxial with the rotating axis.* However, only the first condition is both *necessary **and** sufficient.* Consequently, the axis of the illuminating optics may be tilted with respect to the rotating axis while the initial focal point intersects the rotating axis, sustaining radial symmetry. This important option is illustrated in Figures 4.5 and 4.6. [See "Appendix B: Circular Locus Theorem" in "Fundamental architecture of optical scanning system (Beiser, 1995).

---

* See also Section 3.4.1 for operation in radial symmetry with collimated light parallel to the axis, forming preobjective scanning of Figure 3.6.

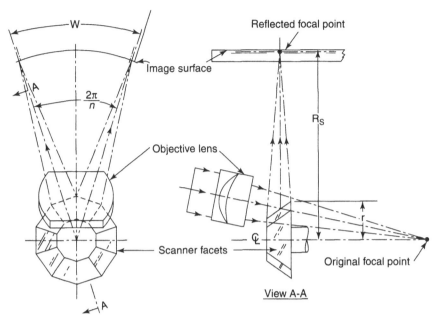

Fig. 4.5   Tilted-axis postobjective pyramidal scanner. Initial focal point on rotating axis provides radial symmetry. On-axis objective lens minimizes its size and aberrations. Two-facet overfilling maximizes the N.A. and scan duty cycle.

Figure 4.5 illustrates the tilted axis pyramidal scanner, drawn to scale for one design. Its scanned focal point (reflected from an over-filled uniformly illuminated keystone aperture, Section 3.2.1.1) traverses a 5-in. width on curved film. At 633 nm, this f/6 cone provides high MTF at 100 lp/mm (5-μm spot size) at a bandwidth of 50 mHz. Figure 4.6 shows its application in recording "cupped" film (with transitions to "flat" on film spools at both ends). The film is maintained with concentric cylindrical curvature in the scanned region by a slotted arced guide (not shown). This basic system was deployed in a family of renowned phototransmission system programs [described briefly in Appendix 1 of *Holographic Scanning* (Beiser, 1988) and extended for wideband data storage and retrieval with added precise tracking of the scanned data.

In addition to allowing use of an objective lens of reduced size and cost, the principal advantages of this tilted axis configuration are—

(1) The f-number of the input objective lens is relaxed (allowed to increase) significantly, by a factor related to $\sin \pi/n$. For $n = 10$ facets, the objective lens f-number increases by a factor of $\approx 3\frac{1}{4}$ times. In the challenging high-speed scanning of an f/6 cone

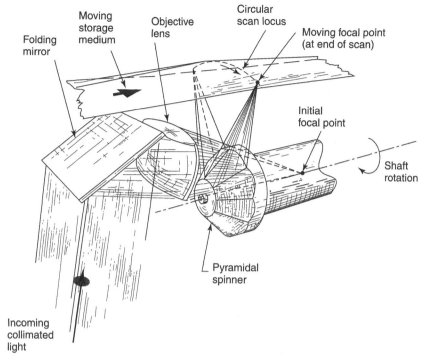

Fig. 4.6 Application of tilted-axis postobjective configuration of Fig. 4.5. "Cupped" film is scanned in circular arc at very high resolution and speed. From "Laser scanning systems," by L. Beiser in *Laser Applications*, Volume 2, by Monte Ross, ©1974, Elsevier Science (USA), reproduced by permission of the publisher.

described above, and allowing for overfilling to uniformize the aperture, this permits design of an easily realizable objective lens.

(2) The objective lens operates in its essentially stigmatic axial region, rather than from a peripheral segment that is subject to to off-axis aberrations.

Although arced scan loci are typical in postobjective operation, significant work has been conducted [Wal] to approach a linearized portion of scan during such operation for special applications. One such development was for material surface processing utilizing a 5-kW CO_2 laser. This linearizing technique balances the intrinsic arced scan with a compensating function utilizing the facet "shift" (end of Section 4.3.5.2) to contribute a complementary arc to straighten the scan locus over an adequate region of operation. Significant value is reported, as applied above. Its use in more general applications is limited to those cases that (a) require relatively narrow scan angles and field widths,

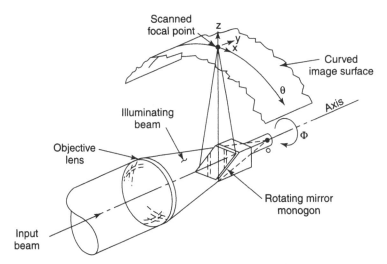

Fig. 4.7 Postobjective scanning by a monogon illustrated as in Fig. 1.10, except for its rectangular aperture. When overfilled uniformly, it develops a sinc²(x,y) PSF, providing identifiable angular rotation during scan. From [Bei3].

(b) tolerate a polygon that is very large compared to the scan width, and (c) where interferance between the scan plane and the input beam can be avoided. This conflict was overcome in the above application by separating the input and output beam planes angularly as in Figures 1.4 and 4.3, although not benefiting from reduction of the angle by double pass. Consequently, as burdened further because of the short focal length (short level arm) of the output beam, the resulting scan bow must be considered in light of the intended application.

4.3.5.5 Image Rotation and Derotation The rotation of an isotropic point spread function (PSF) about its optical axis is nominally undetectable. However, if the illuminating beam forms a PSF that is nonisotropic geometrically or polarized, or if an array or group of points is to be scanned, some arrangements can cause rotation of the images about the projected axis (the principal ray of the scanned beam or group.) For a nonrotation condition, consider the postobjective monogon scanner of Figure 4.7, which is illuminated coaxially with a grossly overfilled beam. The rectangular boundary of the mirror delimits the reflected beam to exhibit a substantially uniform near-field intensity distribution. During mirror rotation, this rectangular near-field cross section maintains the same relationship to the Φ- or Θ-direction. And the projected scanned focused beam also maintains a fixed $(x-y)$ orientation on the image surface. Thus, with the mirror boundary delimiting the illuminating beam, the focal point (having a diffraction-limited $\mathrm{sinc}^2 x,y$ form of PSF) maintains the same relationship to the $(x-y)$ coordinates on the image surface of Figure 4.7. That is, it does not rotate.

If, however, the same scanner is *underfilled* with, for example, an elliptical cross section gaussian beam, this smaller input beam (not the mirror boundary) establishes the elliptical contour of the reflected beam. Here, the focused elliptical Gaussian spot rotates about its projected axis directly with mirror rotation [Gin,Lev]. To see this, consider the major axis of the iluminating ellipse to be vertical, along the z-axis of the center-scan position. When the rotating mirror of Figure 4.7 is in this position (as illustrated), the major axis of the near-field *reflected beam* falls along the y-axis. (The focused Gaussian ellipse, Fourier transformed in quadrature, arrives with its major axis along the x-axis.) When the scanner is rotated through $90°$, the major axis of the incident illuminating beam remains along the z-axis, while the beam encounters the rotated mirror such that the major axis of the *reflected beam* near-field now appears *along the x-axis*. (This transforms to a focused elliptical spot whose major axis forms along the y-axis.) The reflected beam is *rotated about its principal z-axis by the same angle* as the rotation of the scanner. Similarly, if the scanner is illuminated with polarized or multiple beams to form an array of scanned spots, the imaged polarization or array angle will rotate about its z-axis directly with the mirror. This effect is identical in the pyramidal polygon of Figure 3.6. Each mirror forms a marginal segment of the centered mirror of Figure 4.7, imparting the same rotation about its axis to the reflection of the illuminating beam.

This effect does not occur, however, for the mirror mounted per Figure 4.8 or for the prismatic polygon of Figure 3.6. It is also independent of preobjective or postobjective operation. When the principal rays of the input and scanned beams are in a plane that is normal to the axis of rotation, execution of scan does not alter the image (except for possible aperture vignetting or alteration of the reflection characteristics during scan). In the above cases of radial symmetry (Section 3.4.1), the incident angles remain constant while the focused images rotate. Here (in extreme radial *asymmetry*), the incident angles change while the image develops no rotation. Although mirrored scanners seldom operate between these extremes, holographic scanners can, creating possible complication with, for example, polarization

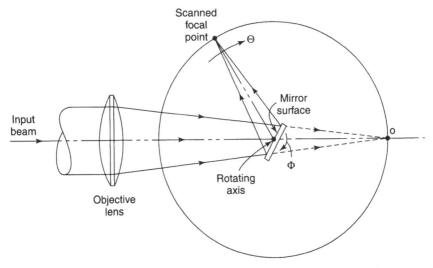

Fig. 4.8   Postobjective mirror with reflective surface on rotating axis (typical galvanometer mount). Scan angle Θ is twice rotational angle Φ ($m = 2$). From [Bei3].

states. This is manifest in the variation in diffraction efficiency of gratings for the p and s polarizations (holographic or otherwise) when they are rotated [Bei1].

To implement derotation, complementary rotation is interposed in the path to cancel rotation caused by the scanner. The characteristic of a coaxial image rotator is that it inverts an image [Lev]. This results in two rotations of the image per rotation of the component. Thus, when counterrotated at $\frac{1}{2}$ the speed of the scanner, a rotator such as the Dove prism can provide coaxial image derotation. Whereas the use of the Dove prism is limited preferably to operation in collimated light,* alternatives are not only available for such use but are adaptable to serve in converging or diverging light [Lev,Bro]. They include a *three-mirror* assembly that simulates by reflection the refractive paths of the Dove prism, a *cylindrical/spherical* optical relay, and a *Pechan prism*, a compact arrangement of prisms having reflective surfaces for operation in noncollimated light. To optimize the location of the rotator, consider the discussion in Section 3.3.1 regarding the propagation of angular errors in view of the resolution invariant. To minimize the transfer of paraxial angular errors that might be introduced by fabrication or

---

* Converging or diverging light that refracts at plane surfaces oblique to the optical axis develops spherical aberration, astigmatism, and coma (notably in the Dove prism, having surfaces at substantive obliquity to the optical axis) [Lev, Bro].

mechanical rotation tolerances, the rotator must be located preferably *in advance* of any beam expansion optics required for system operation. Not only does this allow use of smaller aperture rotational optics, but the pointing errors due to imperfect rotation of the system are reduced by a factor equal to the beam-expansion magnification.

4.3.6 Passive Scanning for Remote Sensing

Figures 1.1 and 1.2 at the beginning of this work identify, respectively, active and passive scanning. The distinctions between the two appear in the nomenclature and in the reciprocal optical path directions of a conjugate imaging system. Ray directions propagate from left to right in Figure 1.1 and are reversed in Figure 1.2. On the left in Figure 1.1 appears the single reference point (source), which is analogous to the single photodetector on the left in Figure 1.2. Similarly, the scanned "moving focal point" (or "flying spot") on the right in Figure 1.1 corresponds to the "moving sampling point" on the right in Figure 1.2.

Concentrating now on remote sensing represented by Figure 1.2, the region on the right provides for scanning in object space. Similarly, on the left appears the opportunity for scanning in image space. Because the scanning of infrared radiation (a principal application) is conducted often by the articulation of mirrors or prisms, and multispectral scanning also adapts well to the use of reflective devices, almost all of the mechanisms discussed above are candidates for remote sensing. Related comments at the ends of Sections 1.5.1.1 and 1.5.1.2 are notewothy.

Fundamental forms of one-dimensional (line) scanning are represented by Figures 1.8, 1.9, and 1.10. The second dimension is often provided by quadrature motion (Fig. 3.9) of the sensing vehicle. In these illustrations, ray directions are to be reversed, so that former (active) scan lines become the sampled paths in object space. In Figures 1.8 and 1.9 (with reversed rays) the oscillating or rotating mirrors appear "beyond" the objective lens, in image space, now directing the changing beam angle to a fixed axis A that leads to a fixed detector (not shown). When remote sensing over great distances, the object-side flux arriving at the objective lens is essentially collimated, forming converging beams on the image side to focus on the detectors. In Figure 1.10 (reversing ray directions), the scan mirror now appears in object space, directing the sampled changing-angle beams to a fixed axis through the lens to radiate on a fixed detector (not shown). Here again, the range from a distant object to the scan mirror could be great, prop-

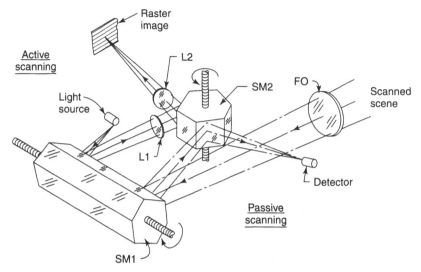

Fig. 4.9 Remote sensing image space scanner and recorder. Radiation from the scene is imaged by the focusing objective lens FO through the vertical and horizontal scanners SM1 and SM2 onto the detector. Simultaneously, a raster image of the scene is formed by modulating a light source with the detected signal and projecting the beam through lenses L1 and L2 and adjacent portions of the same scanners. From [Bar].

agating essentially collimated light to the objective lens to forward the beam to a fixed detector.

An interesting remote sensing system is represented in Figure 4.9 [Bar]. It not only views and detects a sensed scene in two dimensions but with an added light source that is modulated with the detected signal, it doubles up on the scan mechanism to record a two-dimensional raster image, essentially in real time. The passive scanning portion (in the foreground) accepts the substantially collimated radiation from the scene over a changing field angle and conveys it through the focusing objective lens FO to the vertical deflector SM1 and then to the horizontal deflector SM2, to be directed to a fixed detector. This illustration is schematic and not to scale. Lens FO must provide a sufficiently large aperture to include the full field of descanned flux. If lens FO is placed closer to deflector SM1, which in turn is placed closer to deflector SM2, consistent with avoiding interference of the scanned beams, the field overage may be more practical. A good way to evaluate these moves is to consider the detector as a light source, reverse ray directions, and review the consequences of variation of the pupil relief distance, per Section 4.3.5 and subsequent discussion. The active scanning portion (in the background) illustrates some of this consideration by placing lenses L1 and L2 closer to the deflectors. Furthermore, lens

L1 and L2 apertures that are large enough to accept the scanned angles are required.

4.4 HOLOGRAPHIC SCANNERS

Holographic scanners employ many of the disciplines utilized by rotating polygons [Bei1]. Almost all comprise a substrate that is rotated about an axis. An array of diffractive elements (often mastered interferometrically) is disposed about the periphery of the substrate, to serve as transmissive or reflective facets. As with polygons, the number of holograms n is determined, per Equation 4–5, by the desired optical scan angle and duty cycle. And, as expressed for all angular scanners, the resolution N is determined primarily by the scan angle and beam size, per Equations 3–5, 3–11, and 3–13. The similarities relate even more closely in radially symmetric systems (Section 3.4.1), in which the geometric scan functions can be identical to those of the polygon. This is illustrated below. With such similarities, what are the distinctions, advantages, and limitations of hologaphic scanning?

The distinctions may be listed as follows:

1. The substrate surface appears smooth and uniform, exhibiting no facet discontinuities.
2. The holographic (diffractive) facets are microscopically thin grating elements, formed on the otherwise smooth substrate surface.
3. The substrate exhibits ideal rotational symmetry, typically as a flat disk (although not so limited) mounted on a rotating shaft.
4. The scanner may be designed to operate in optical transmission or reflection.
5. The diffractive properties of the holographic facets exhibit spectral and angular selection characteristics.
6. The facet grating may be "linear" (causing only angular displacement) or lenticular (providing added optical power).
7. The orientations of the input and output beams (with respect to the scanner) are design variables.

Consequential advantages may be listed as follows:

1. Replacement of conventional facets with holograms eliminates surface discontinuities. This reduces significantly the aerodynamic loading and windage, allowing more efficient high-speed rotation [Shep, Lenn].

2. Elimination of radial variations also reduces the inertial deformation of the substrate and its facet elements under high-speed rotation [Bei2].

3. When the facets are operated in the Bragg regime (Fig. 3.8), beam misplacement due to shaft wobble is reduced significantly [Kra1,Bei1].

4. Optically exposed facets require no physical contact during exposure, allowing precise shaft and facet indexing and positioning.

5. Accurate replication of the diffractive elements on the substrate (from an optimized master) can reduce the production cost significantly.

6. Beam filtering in retrocollection allows spatial and spectral selection.

7. Facets may be varied (during exposure) in focus, size, and orientation.

Complementing the above advantages are the consequential limitations:

1. Design, fabrication, and test of holographic scanners entails unique technical discipline, along with special instrumentation, metrology, and processing controls. It requres astute orientation in diffractive optics and a significant investment in R&D manpower and facilities.

2. The spectral and angular beam sensitivity of diffractors (distinction nos. 5, 6, and 7) can impose beam shift complications when reconstructing at a wavelength that differs from that of exposure. Also, small wavelength shifts from laser diode drift and/or mode–hopping may appear, potentially perturbing the diffracted output beam angle.

3. Systems that retain radial symmetry for circular scan uniformity may be limited in Bragg angle wobble reduction. They can require auxiliary wobble compensation (Chapter 5), such as anamorphic error correction. Also, for flat-field scan, they must be augmented with a flat-field lens.

4. Systems that do depart from radial symmetry and operate in the Bragg regime can require interrelated balancing of linearity, scan angle range, wobble correction, scan bow correction, radiometric uniformity, and insensitivity to beam polarization [Kra3].

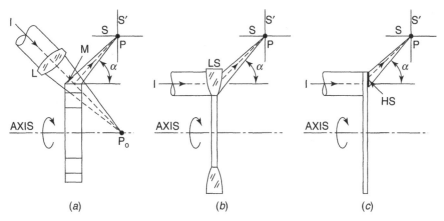

Fig. 4.10 Analogous nonholographic scanners (a) & (b) and holographic scanner (c) form identical radially symmetric outputs through angle α at focal point P. Cylinder S is concentric with the axis, or flat surface S′ is perpendicular to the axis. In (a) mirror M reflects converging input beam I from lens L to point P. In (b) lens segment LS refracts focusing beam to point P. In (c) holographic segment HS diffracts focusing beam to point P. From *Holographic Scanning*, L. Beiser, ©1988 John Wiley & Sons, Inc. Reprinted by permission of John Wiley & Sons, Inc.

4.4.1 Scanner Configurations and Characteristics

To illustrate the analogous relationship of conventional and holographic devices noted in the introductory paragraph, consider Figure 4.10. It shows a transition of three scanning techniques that yields identical scan geometries. All are radially symmetric, scanning a concentric cylindrical information surface S as for "internal drum" application. When the cylindrical medium is unwrapped and flattened, the scans appear as parallel straight lines. An alternate information surface S′ is flat and mounted perpendicular to the rotating axis. When this medium is indexed "vertically," the scans form an array of circular arcs. (This can be useful if, for example, an "original" document is scanned in the same manner to derive a signal for facsimile reproduction.) All three systems yield the same image characteristics. Heuristic Figure 4.10 reresents the transition from (a) a conventional polygon (operating unconventionally) to (b) an uncommon lenticular scanner, to (c) the configuration of a basic form of holographic scanner.

The prismatic polygon at (a) is postobjective. The input bcam I is converged by objective lens O initially to axial point P_0 while overfilling mirror M. Thus operating radially symmetric, the beam reflected from M through angle α scans focal point P in a circular arc on the inside of cylindrical surface S. The device at (b), with marginal lens segments LS, executes the same scan function. Overfilled segment LS redirects collimated paraxial input beam I through angle α to focus at P. Upon rotation of the array, it also forms circular arcs on the inside of surface S. At (c) is one basic form of holographic disk scanner. The input beam I is also paraxial, and the overfilled holographic sector HS serves as the lens segment LS to redirect and focus I through angle α to scanned point P on surface S. In this example, holographic segment HS exhibits lenticular power to emulate lens segment LS. It may also be considered a marginal segment of an infinity conjugate zone lens, composed of a concentric array of alternating opaque and transparent circles. The zone lens is developed comprehensively [Bei1] from this infinity conjugate form through the finite conjugate form to the generalized form, as relating to the formation of the holographic grating and its periodicity.

The choice of holographic grating contour is a major design consideration. The two principal options are (per item 6 in "distinction" listing above) "linear" and "lenticular." The linear form is a plane diffraction grating having equally spaced straight grating elements. It will be seen (Section 4.4.1.1) that this structure is in many respects analogous to a refractive wedge prism. The lenticular form is analogous to a lens element, as represented in (b) and (c) of Figure 4.10 and its zone lens prototype. Although it may appear that the lenticular grating is also to be disposed on a flat surface, this is not a limitation to valid operation (as indicated by item 3 in "distinctions" listing). In fact, lenticular holograms of notable holographic scanners were formed on curved substrate surfaces [Bei1]. Several such forms are illustrated and discussed below.

With regard to the plane linear transmission grating, Figure 3.8 illustrates its typical application. The equal input and output angles β_i and β_o (Bragg condition at the center of disk rotation) provide the unique property of reducing significantly the degrading effect of disk wobble $\Delta\alpha$. This would otherwise vary the output beam angle (with respect to the facet normal), causing nonuniformity of scan line spacing. Analysis [Kra2, Bei1], starting with the classic grating equation represented above by Equation 3–14 and repeated here

$$\sin\beta_i + \sin\beta_o = \lambda/d \qquad (4\text{--}14)$$

In which d is the grating spacing, reveals a relationship for the tilt error in the vicinity of Bragg operation for both transmission and reflection gratings. This differential in output beam angle $d\beta_o$ for a differential tilt in hologram angle $d\alpha$ within an angular error $\Delta\alpha$ is given by

$$d\beta_o = \pm \left[1 \mp \frac{\cos(\beta_i + \Delta\alpha)}{\cos(\beta_o \mp \Delta\alpha)} \right] d\alpha \qquad (4\text{--}15)$$

in which the upper sign is for transmission and the lower for reflection gratings.

For a typically small $\Delta\alpha$ error, the reduction of the effect of wobble in transmission gratings is apparent from Equation 4–15, where the fractional term is approximately equal to 1, leaving (for $\beta_{ot} \equiv$ output angle in transmission)

$$d\beta_{ot}/d\alpha \rightarrow 0 \qquad (4\text{--}16)$$

Numerically, for $\beta_i = \beta_o = 45°$, a substantive angular error of 0.1° (360 arc Sec.) is reduced to 1.3 arc Sec.—a 277× improvement. For the reflective grating, the positive sign before the fraction applies, representing an addition of unity (1) to the existing (1), yielding (for $d\beta_{or} \equiv$ output angle in reflection)

$$d\beta_{or}/d\alpha \rightarrow 2 \qquad (4\text{--}17)$$

This is similar to the doubling effect of a polygon ($m = 2$) in mirror reflection. Thus a transmission grating operating in the Bragg regime ameliorates the effect of such wobble error of a typical reflective facet polygon by a factor of approximately 550 times. Although this indicates an extremely attractive wobble reduction factor compared to that of the polygon, this is achieved only in the center of scan. Furthermore, this condition is also accompanied by a nonminimized scan line bow. However, by unbalancing the nominal Bragg angle condition of equal input and output angles to one having a small differential ($\Theta_i - \Theta_o \simeq 1.41°$) [Kra3], the wobble correction is equalized at the center and ends of scan while minimizing the scan line bow. Under these conditions, the effect of scanner wobble is reduced by a factor of approximately 82 times compared to that developed by a typical prismatic polygon.

With regard to the resolution of holographic scanners, as noted in the introductory paragraph to Section 4.4, this factor is governed by the same equations as for conventional scanners. When the resolution is augmented with finite values of r and f (Equations 3–11 or 3–13), the value of scan magnification m (see Sections 3.4.1 and 3.4.2) is represented, interestingly, by Equations 3–14 or 4–14, the grating equation.

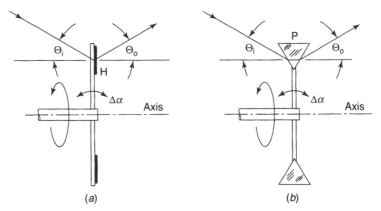

Fig. 4.11   Analogous Bragg diffraction and prism refraction at minimum deviation. [a] Holographic scanner. Hologram H operating in Bragg regime. [b] Analogous prismatic scanner. Prism P at minimum deviation. From *Holographic Scanning*, L. Beiser, ©1988 John Wiley & Sons, Inc. Reprinted by permission of John Wiley & Sons, Inc.

4.4.1.1   Analogous Relationship of Bragg Diffraction and Wedge Prism Refraction

The discussion above of the reduction of the effect of wobble by the linear plane grating identified an analogous relationship to the refractive wedge prism. Pursuit of this characteristic provides a revealing expression of this effect. Consider Figure 4.11 depicting the two configurations. View *a* is a repeat of Figure 3.8, with notations of the Bragg angles β repaced with the more generic notations Θ, to be consistent with those of a hypothetical wedge prism scanner of view *b*.

As may be seen by a comparison of classic optics presentation [J&W] of effective prism tipping and the analysis [Kra1,Bei1] of hologram tipping, a strong parallel appears between the effects of small departures from the Bragg condition and departure from the minimum deviation of a prism. Notably, for a prism (in air) of apex angle 60° and glass index of 1.5, compared with a linear hologram having $\Theta_i = \Theta_o = 45°$, the plotted curves are almost identical for deviations of ± a few degrees. Keeping in mind that a deviation of ±1° is a "gigantic" number of 3600 arc seconds for the wobble of a rotating system, which is seldom allowed to exceed 60 arc seconds, the comparison is significant.

For the analysis of the prism system, a ray-trace modification of Figure 4.11*b* includes identifying the prism with refractive index *n* and apex angle α, constructing the normals to the input and output faces, and noting the input ray angle to its face normal as Θ_{ip}, the output ray angle to its face normal as Θ_{op}, the input angle of the ray propagating within the prism as α_i, and the output angle of the ray propagating within the prism as α_o. Then, expressing Snell's law at both faces,

$$\sin\Theta_{ip} = n\sin\alpha_i \qquad (4\text{--}18a)$$

and

$$\sin\Theta_{op} = n\sin\alpha_o \qquad (4\text{--}18b)$$

The internal angles of the prism can be shown to be related to its apex angle as

$$\alpha_i + \alpha_o = \alpha \qquad (4\text{--}19a)$$

By operational symmetry,

$$\alpha_i \simeq \alpha_o = \alpha/2 \qquad (4\text{--}19b)$$

yielding,

$$\sin\Theta_{ip} \simeq n\sin\alpha/2 \qquad (4\text{--}20a)$$

and

$$\sin\Theta_{op} \simeq n\sin\alpha/2 \qquad (4\text{--}20b)$$

When written in the same form as the grating equation (see Equation 4–14),

$$\sin\Theta_{ip} + \sin\Theta_{op} \simeq 2n\sin\alpha/2 \qquad (4\text{--}21)$$

whence, for $\sin\alpha/2 \simeq \alpha/2$,

$$\sin\Theta_{ip} + \sin\Theta_{op} \simeq n\alpha \qquad (4\text{--}22)$$

Equation 4–22 is seen to be functionally the same as Equation 4–14 with

$$n\alpha \Leftrightarrow \lambda/d \qquad (4\text{--}23)$$

representing operationally equivalent constants. The product of the refractive index n and the apex angle α of the prism is equivalent to the ratio of the wavelength λ to the grating spacing d of the diffractor. Furthermore, their typical working values each form magnitudes of approximate unity. Thus, following the procedure that yielded Equation 4–15 and its low wobble consequences for the linear holographic scanner, the same inferences may be drawn for a hypothetical scanner utilizing prisms oriented in minimum deviation. Because the operation of such a prism scanner could be burdened by nonuniformities of prism assembly, rotational imbalance, and excessive inertial stress and strain, this analogous configuration is offered for its unifying pedagogic value, rather than for its practical utilization.

4.4.2 Implementation of Holographic Scanners

It is informative to consider some earlier holographic scanners and their relationship to to contemporary designs. An example of one of the first systems is illustrated in Figure 4.12, representing the family of Holofacet scanners [Bei1]. The most prominent of that group (the first patented holographic scanner) was tested in the early 1970s to the highest resolution and speed: $N = 20,000$ elements/scan at 200 Mpixels/s. Designed for photoreconnaissance imaging, that apparatus

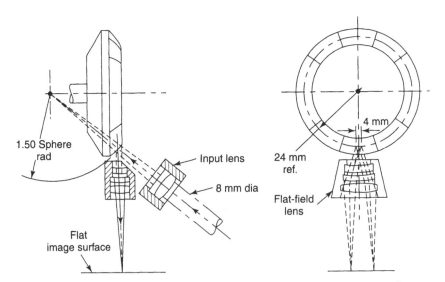

Fig. 4.12 Reflective Holofacet scanner, underfilled. Input lens converges beam on rotating axis. Output scanned collimated beam is focused by flat-field lens upon flat image surface. Applied as microimage recorder (100 lp/mm over 11-mm format). From *Holographic Scanning*, L. Beiser, ©1988 John Wiley & Sons, Inc. Reprinted by permission of John Wiley & Sons, Inc.

is now in the permanent collection of the Smithsonian Institution. The Figure 4.12 version also emulates a reflective pyramidal polygon with the input beam converging to focus on the axis (see Fig. 4.5). It allows, however, formation of a *flat field*. This option of reconstructing a collimated beam for entry into a flat-field lens is available uniquely with holographic facets when the holograms are exposed with a collimated "object" beam (and a required "reference" beam.) No other scanning technique (reflective, acoustooptic, electrooptic, phased array, etc.) provides for arbitrary formation of the output beam configuration. For reference, flat-field operation of a pyramidal polygon is represented in Figure 3.6, in which the polygon is illuminated with a collimated beam parallel to the axis. This illumination option for operation in radial symmetry is also available in the holographic systems described below.

The Figure 4.12 configuration was designed for microimage recording at 100 lp/mm over an 11-mm flat image surface at high speed. This corresponds to 2200 spots of 5-μm width over the 11-mm format. If imaged with a lens of increased focal length, as for reprographics application, this forms 260 DPI (dots per inch) over an $8\frac{1}{2}$-in. format. As is typical where laser power is to be conserved, the facets are underfilled

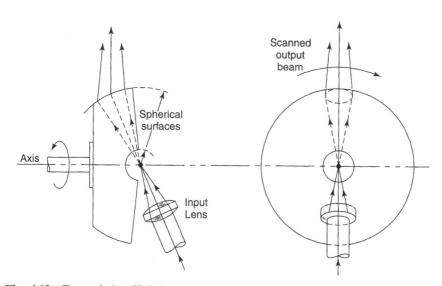

Fig. 4.13 Transmissive Holofacet scanner utilizing solid glass substrate. Provides large-aperture scanned beam for high resolution. Solid substate maintains stability and surface integrity at higher speeds than does the thin-walled design of Fig. 4.14. From *Holographic Scanning*, L. Beiser, ©1988 John Wiley & Sons, Inc. Reprinted by permission of John Wiley & Sons, Inc.

and sufficiently wide to provide a high duty cycle (see Section 4.3.2). The very high-resolution and high-speed reconnaissance configurations utilized the important option of *overfilling* the facets (see Section 4.3.3). This is most notable in the system offering the highest single-channel performance, 50,000 elements/scan at 500 Mpixels/s; the Ultraimage Cylindrical Holofacet Scanner [Bei1]. In contrast to a laser printer forming 3000 elements/scan at 10 Mpixels/s, it provides effectively montaged images of 50 such printers.

Although the above-described scanners were designed to diffract in reflection, as formed on a solid beryllium substrate (to maximize inertial stability at high speed), they may be formed on transparent substrates to operate in transmission at appropriate lower speeds. Two such

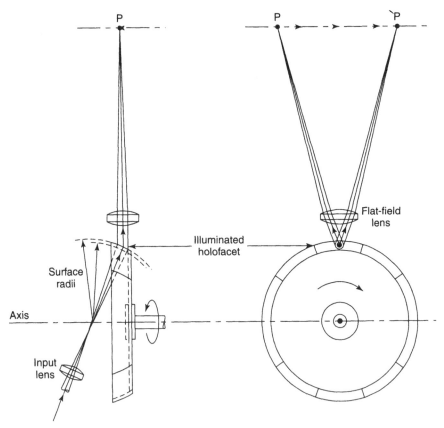

Fig. 4.14 Transmissive Holofacet scanner for business graphics applications. Underfilled facets direct collimated output beam through flat-field lens, forming straight line scan locus P. From *Holographic Scanning*, L. Beiser, ©1988 John Wiley & Sons, Inc. Reprinted by permission of John Wiley & Sons, Inc.

arrangements appear in Figures 4.13 and 4.14 [Bei1]. Both are radially symmetric, with the input beam focused on the rotating axis. Figure 4.13 shows the use of a thick glass substrate to provide internal elastic constraint during higher-speed rotation. In Figure 4.14, the holograms are formed on a thin glass substrate that tolerates the lower-speed operation. The Figure 4.13 design forms a circular scan locus per Figures 4.5 and 4.6, whereas that of Figure 4.14 develops a flat field for reprographics, typical of business graphics scanning and recording.

A transmissive holographic scanner for reprographics was formed on a cylindrical glass substrate, as described in 1975 [P&W] and illustrated in Figure 4.15. Serving as a document scanner, it collects a portion of the scanned radiation that is backscattered from the document. This, in turn, is rediffracted by the "Hololens" and effectively "descanned" to converge toward point O on the rotating axis. The now fixed converging beam is intercepted by the mirror and reflected toward a small detector that is fixed at the reflected focal point O' to develop the scanned signal. This system utilizes reciprocal diffractive properties, whereby reillumination of the hologram by the negative of the "object" wave reconstructs the negative of the "reference" wave. Because this is true for any hologram, it is an option for any holographic scanner configuration. It is noteworthy that retroreflected signal detection (retrocollection) is not unique to holographic scanning (Section 1.2.2.1). Such reciprocity properties may serve almost any optical scanner, so long as the acceptance aperture subtends sufficient return signal to override systematic noise. Diffractive elements can also provide spectral filtering.

Early in Section 4.4.1 there is a discussion of one of the most significant advances in holographic scanning, that is, the utilization of a rotating transparent disk having an array of linear plane grating "facets" disposed along its peripheral radius and operating in the Bragg regime. The principal beam configuration is illustrated in Figures 3.8 and 4.11*a*. This arrangement, which became a model for contemporary design, was initiated in the early 1980s [Kra1,Kra2,Bei1]. Although holaphic scanning allows the options of many variational configurations, utilizing transmissive or reflective substrates, having linear or lenticular gratings, maintaining rotational symmetry or asymmetry, or formed interferometrically or by computer pattern generation, the most popular format is the above transmissive disk disposed with a peripheral array of linear grating "facets" operating in the Bragg regime. Although the output beam is uniquely not normal to the rotating axis, the scanned beam is constrained within a plane with reasonable integrity over a useful scan angle. Thus, with typical collimated input

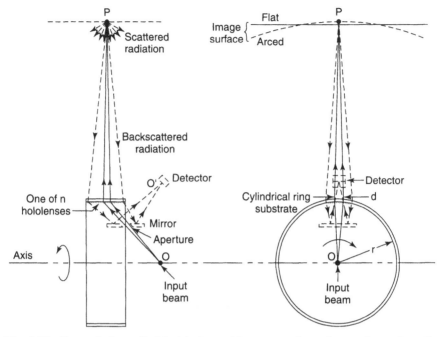

Fig. 4.15 Transmissive cylindrical holographic scanner. Input beam, focused on the rotating axis at O, expands and underfills a hololens that diffracts the beam to focus on P along the arced image surface. Returning dashed lines designate optional collection of backscattered radiation for deriving a document scanning signal at the detector at O′. From *Holographic Scanning*, L. Beiser, © 1988 John Wiley & Sons, Inc. Reprinted with permission of John Wiley & Sons, Inc.

and output beams, a flat-field lens serves to focus the scanned beam along a straight line.

The first operational form utilized the analytically optimal Bragg angle of 45° ($\beta_i = \beta_o \simeq 45°$ in Fig. 3.8) to yield a sufficiently straight line over a large enough scan angle to be useful. This Bragg orientation, per grating Equation 3–14 yields scan magnification $m = \lambda/d = \sqrt{2}$, forming the output scan angle, which is $\sqrt{2}$ larger than the disk rotation angle. However, two restrictions are imposed:

1. High diffraction efficiency from relief gratings (e.g., photoresist) requires high depth-to-spacing ratios, whereas the spacing $d = \lambda/\sqrt{2}$ must be extremely narrow. This is not only difficult to achieve holographically, but also difficult to replicate.
2. Such gratings exhibit a high polarization selectivity, imposing a variation in diffraction efficiency during scan when illuminated with a polarized input beam.

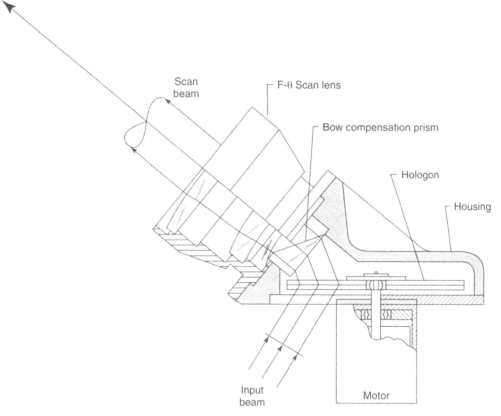

Fig. 4.16 Plane linear grating (Hologon) holographic disc scanner. 30° Bragg angle provides more fabricatable and polarization-insensitive grating structure. Requires, however, bow compensaton prism to maintain scan beam in plane perpendicular to paper. From [Bei8].

These limitations are accomodated by reducing the Bragg angle and by introducing (in one method) a bow compensation prism to straighten the scan line. Figure 4.16 illustrates such a system: a high-performance "Hologon" scanner designed for application in the graphic arts. With the Bragg angle reduced to 30°, the magnification (Equation 3-14) is reduced to $m = \lambda/d = 1$ (as in radially symmetric systems). This increases d to equal λ for a more realizable grating depth to render a higher diffraction efficiency. It also reduces significantly the angular sensitivity of the grating to input beam polarization.

In less demanding tasks than the above graphic arts "typesetting" function, (e.g., laser printing), the elegance of the original 45° Bragg configuration has been adapted to achieve self-focusing, as illustrated in Figure 4.17. An added holographic lens complements the wavelength

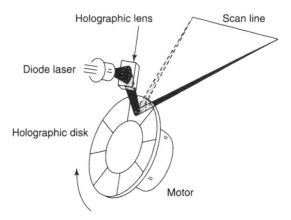

Fig. 4.17 Hologaphic disk scanner with corrective holographic lens that balances the cross-scan error due to wavelength shift and adds optical power to focus the scanned beam. 45° Bragg angle utilizes no bow compensation. Focusing holographic facets imposes precise centering of scanner disk. From L. Beiser, *Laser Scanning Notebook*, SPIE Press (1992). Reproduced by permission of the publisher.

shift of the diode laser and shapes the laser output for illumination of the scanner [Kay,Ike,Yam]. However, the control of such multifunction systems is compounded by the balancing of characteristics required to achieve their objectives. Notable is the critical orientation of the lenticular holograms on the disk and of the centration of the rotating disk [Bei1] to approach repeatedly precise positioning of the scan lines.

A system requiring significant analytic evaluation to optimize performance of several interrelated elements is represented in Figure 4.18, the main diagram of the 17 figures in its patent [C&R]. Although this illustration appears to show the path of a typical beam of finite width, it actually represents the principal rays of three beams of slightly different wavelength, as derived from a (drifting or mode-shifting) diode laser. Lines P_1, P_2, and P_3 trace different positions (displacement highly exaggerated) of three such instances, where $\lambda_3 > \lambda_2 > \lambda_1$ (Typ. ±1 nm). Corrector hologram 108 diffracts a beam of longer wavelength to a shorter radius at scanner disk 103, such that the scanned angle and resulting line width remain constant; the opposite correction is for a shorter wavelength. However, because hologram 108 is oriented differently than in Figure 4.17, nonparallel to the scanner disk [Kay], it *increases* the cross-scan error. This, in turn, is compensated by cross-scan corrector hologram 101. Cylindrical element 105 is an alternate component to correct the astigmatism introduced by hologram 101. Along with cylindrically curved mirror 107, all these components inter-

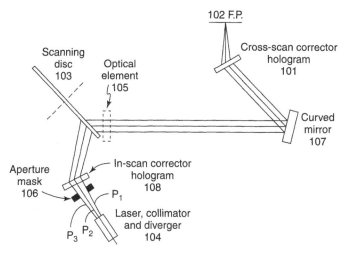

Fig. 4.18 Integrated holographic scanner compensates for diode laser wavelength shifts for both along-scan and cross-scan directions. Rays P_1, P_2, and P_3 at three different wavelengths are restored by hologram 108 to provide uniform scan line lengths and by hologram 101 to form a common focal line. Utilizing no refractive flat-field lens, curved mirror 107 forms telecentric output and corrects scan bow. Interactions complicate analysis and design. Achieves use of replicated holograpic scanner disks. From [C&R].

relate and cooperate to form a telecentric focused line over its 9-in. scanned width.

This holographic disk operates with a Bragg angle of approximately 30°. Recalling the consequences of the Figure 4.16 Hologon system, which benefits from the favorable characteristics of this Bragg angle (unity magnification and more replicable grating contour), unfortunately, scan bow is introduced. Whereas that scanner utilizes an added prism to complement the bow, this system employs a tilted-axis orientation of the curved mirror to generate a complementary bow, furthering the interrelation of components. With production economy in mind, the developer of this system devoted extensive research in injection and cold-form molding and fabrication techniques [Row]. Utilizing the above-discussed grating replicability feature on the master holograms, they succeeded in the quantity reproduction of a variety of precision holographic elements and scanner disks.

Many additional holographic scanners have been implemented. Notable are two innovative adaptations [Kra3] of the more familiar pyramidal polygons. Because they retain the radial symmetry of the polygon types, the scan beams can propagate perpendicular to the rotating axis and develop perfectly straight scan lines. The first is effec-tively a transformation of Figure 1.9, in which the 45° mirror is replaced with a complementary-angled 45° plane linear grating hologram. Thus, with typical collimated input illumination along the rotating axis, the diffracted beam scans through a flat-field lens, as in Figure 1.9. Because the Bragg angle is 45°, this scanner exhibits the optimal effects of reduced wobble coupled, however, with the two limitations identified above of challenging grating dimensions and polarization sensitivity. The second configuration is a multifacet extension of the first, essen-tially an analog of Figure 3.6, underfilled. It is comprised of four in-dividual (plane linear grating) holographic facets mounted at 45° in a cylindrical carrier, with the pyramidal apex facing away from (rather than toward) the input beam. The facets form a true four-sided pyramid. As such, they exhibit the radial discontinuities of a polygon scanner (e.g., differential inertial stess). And, assembled individually as completed holograms, they require orientation and tilt adjustment to attain accurate scan line repeatability. Alternatively, the sensitized plates may be mounted for exposure in an armature having identical bearing support. Indexed angularly, exposed and processed in situ, the armature is then assembled into the scanner, holograms well oriented. A similar self-aligning technique was employed more easily on the con-tinuous substrates, for example, spherical per Figure 4.12, and disk of Figures 4.16 and 4.18. *Holographic Scanning* (Beiser, 1988) provides a comprehensive analytic and historic review of much additional holo-graphic scanner development.

4.5 OSCILLATORY (VIBRATIONAL) SCANNERS

The scan nonuniformities that can arise from minute errors of rotating mirrored or holographic facets may be avoided by eliminating all but one facet. This changes a polygon to a monogon, which adapts well (per Fig. 4.7) to internal drum scanning. It achieves a very high duty cycle when executing a large scan angle within a cylindrical image surface. The same scan, however, when projected through a flat-field lens per Figure 1.8 or 1.9, allows only a limited scan angle, resulting in a limited duty cycle from a monogon. If the mirror is oscillated rather than rotated through a full cycle, the wasted scan interval may be reduced. This component must, however, satisfy system speed, resolution, and linearity. Such devices include the familiar galvanometer and resonant scanners [Ayl, Bei2, Mon1, Rei] and the less frequently encountered piezoelectrically driven mirror transducers [Bei2,Rei] and the "fast-steering" larger mirrors driven by push-pull pairs of voice-coil type actuators [Ber,New,Swe].

The typical mirror mounting position places its reflecting surface coincident with the rotating axis (Fig. 1.8). In this orientation, with the plane of the input and output beams perpendicular to the rotating axis, the optical scan angle Θ is twice that of the rotation angle Φ (magnification $m = 2$). In rare, although useful alternate instances, the mirror reflecting surface is positioned nominally at 45° to the rotating axis (Fig. 1.9). In this condition, with the illuminating beam coaxial to (or derived from focus on) the rotating axis, the optical scan angle Θ is equal to the mechanical rotation angle Φ, exemplifying radial symmetry. Along with some distinctions in performance (see, e.g., Section 4.3.5.5 re image rotation), the effects on scanner-lens relationships are discussed in Section 4.6.

4.5.1 The Galvanometric Scanner

Historically, the galvanometer scanner evolved from the famous D'Arsonval meter movement having a "moving coil" armature suspended within a strong permanent magnet field [Bei2, Fig. 27]. It has appeared in three forms [Mon1]. A modified multiturn moving coil armature and solid cores of either iron or magnetic material to provide increased torque. The moving magnet transducers generally render the highest resonant frequencies and the highest scan speeds.

As shown in Figure 4.19a, the typical contemporary moving magnet galvanometer transducer is similar to a torque motor. Permanent magnets in the stator provide a fixed field that is augmented (±) by the variable field developed by the control current through the stator coils. Seeking a new balanced field, the rotor (and mirror) executes a limited angular excursion (±Φ/2) that is restrained by an elastic suspension system. The galvanometer is a broadband device, damped sufficiently to operate over a wide range of frequencies, from zero to an upper value related to its mechanical resonance. Thus it can provide the basic sawtooth waveform for raster formation, having a long active linearized portion and a short retrace interval of time τ. Figure 4.20 (solid line) shows a typical galvanometer sawtooth scan. Of the total cycle period T, a useful linear ramp portion occupies $0.7\,T$ (70% duty cycle). At moderate frequencies (~100 Hz), duty cycles >85% have been achicvcd [Gad]. Also, as a broadband device, it can provide random access, positioning the beam rapidly to an arbitrary location within the time τ. For this important feature of waveform shaping, the galvanometer was early catagorized [Bei2] as a *low-inertia scanner* (Fig. 4.1). Ex-

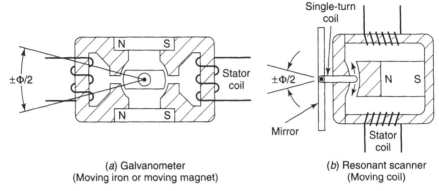

(a) Galvanometer
(Moving iron or moving magnet)

(b) Resonant scanner
(Moving coil)

Fig. 4.19 Examples of galvanometer and resonant scanner transducers. Permanent magnets provide fixed field, augmented by variable field from controlled current through stator coils. [a] *Galvanometer*: Torque rotates iron or magnetic core ±Φ/2. Mirror surface (not shown) on shaft axis. [b] *Resonant scanner*: Torque induced into single-turn armature coil (looped in plane perpendicular to paper) rotates the mirror ±Φ/2 suspended typically within torsion bars. One stator coil may be nondriven and supply induced current for velocity pick-off. From [Bei7] Wiley-VCH (1995) *Encyclopedia of Applied Physics* Vol. 12. Fig. 18 on p. 352.

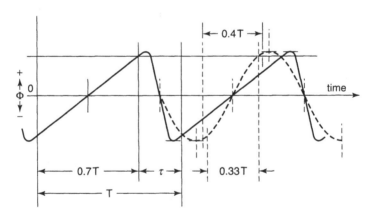

Fig. 4.20 Waveforms (Φ vs. time) of two vibrational scanners having same periods and zero crossings. *Solid line*, galvanometer with linearized scan at 70% duty cycle. *Dashed line*, resonant scanner providing 33.3% duty cycle (unidirectional) with 2:1 slope change, or 40% duty cycle with 3.24:1 slope change. See text for further data. From L. Beiser, *Laser Scanning Notebook*, SPIE Press (1992). Reproduced by permission of the publisher.

cellent reviews of galvanometer design have been published recently [Mon2,Ayl].

4.5.2   The Resonant Scanner

A resonant scanner transducer is illustrated in Fig. 4.19*b*, a single-turn coil (within its controlling magnetic field) coupled directly to the mirror [Mon]. Other forms utilize armatures of iron or magnetic materials [Rei,Tuc]. With damping removed almost completely, large angular oscillations can be sustained only very near the resonant frequency of the armature and its suspension system. The resonant scanner is thus identified with near-perfect sinusoidal oscillations at a fixed and typically high frequency. Figure 4.20 (dashed lines) illustrates such a sinusoid having the same zero crossings as those of the (solid line) galvanometer sawtooth function. Contrary to its prevalent designation as "low inertia," the resonant scanner acts as a pendulum, providing rigid time increments, as though it exhibited high inertia. In some designs, a small amount of inertial tuning can be provided to trim the resonant frequency slightly. Although the rotary inertia of the armature system is low, allowing high cyclic rates, it permits no random access and no waveform shaping, as do the galvanometer, acoustooptic, electrooptic, and other wideband scanners designated as low-inertia devices. Thus, in the Figure 4.1 categorization of optical scanning techniques, the "oscillatory resonant" scanner is classified as "high inertia."

   With its high mechanical Q (very low damping), the resonant scanner provides only harmonic scans. Because we typically seek a linearized scan segment, significant adaptation of the sinusoid is often required to render scan utility. As illustrated in Figure 4.20 (dashed lines), we need to utilize a central portion of the sine function that is sufficiently linear to linearize the image further by timing the pixels, as by extracting them out of memory at a complementary rate [Twe]. Some numeric values are useful indicators of the degree of correction required. To limit the variation in pixel rate to 2:1 (velocity at crossover to be twice that at the selected scan limit), one may restrict the angular scan to 60°/90° = 66.7% of its peak excursion. When scanning with only one slope of the sinusoid (as for raster formation on a uniform medium transport), this represents a duty cycle of only 33.3%. To raise the duty cycle, one must accomodate a greater variation in data rate. If, for example, we limit the useful scan to 80% of its excursion (40% duty cycle; one slope), the velocity variation rises to 3.24x. That is, the data rate at crossover is 3.24 times that at the scan limit. Accompanying this scan velocity variation is the corresponding variation in dwell time of the pixels, resulting in loss of pixel exposure or detectivity: 2:1 for 33.3% duty cycle and 3.24:1 for 40% duty cycle. This, too, may require compensation over the full scan interval, utilizing position-dependent sensitivity controls [Mon,Rei,Twe,M&G] to provide complementary pixel modulation. In contrast, the broadband galvanometer can render a linarized scan at much higher duty cycles. At a duty cycle of 70%, its data rate need be only 1.43x that for an ideal duty cycle of 100% (Section 4.3.3).

Also, its data rate at crossover is increased significantly over that of the galvanometer. For equal peak-to-peak excursions (Fig. 4.20), the bandwidth is approximately $2\frac{1}{2}$ times that of the galvanometer (with 70% duty cycle), as represented by their relative slopes at crossover. For equal angular excursions of both, with the resonant scanner operating at, for example, 40% and 33.3% duty cycle, its bandwidth at crossover rises by a factor of approximately 3.1 and 3.7 times that of the galvanometer, respectively. With thoughtful provision for the above comparative factors (in electronic and system control), the resonant scanner renders application in compact linearized scanning tasks, including accomodation for operation in the forward and return scan directions, for increased duty cycle.

4.5.3 Suspension Systems and Position Control

In the galvanometer and resonant scanners, the bearings and suspension systems are the principal determinants of detailed scan uniformity. A given radial run-out (of a ball bearing, for example) will cause more shaft wobble the shorter the suspended distance between bearings. Thus the shaft length and mass are critical design factors. Sufficient stiffness and length of the armature are required to restrain cross-scan wobble. However, to maximize the upper oscillating frequency, the armature is restricted in size and mass. Fortunately, the reciprocating armature suspended within such bearings tends to repeat its path (and its perturbations) faithfully over many oscillations, making adjacent scans of these devices more uniform than if the same shaft rotated continuously within the bearings, as in a motor.

Some bearings are elastic flexure, torsional, or taut-band devices that are mounted rigidly at their ends and contribute almost no scan perturbations. Because of the high mechanical Q of these suspensions, they are most often applied to the resonant scanner [Tuc]. When properly damped, they can also serve for galvanometer scanners, suffering a small reduction in peak-to-peak excursion but gaining in more uniform scan with very low noise and extended life. Some considerations are their low radial stiffness and possible cross-coupling of perturbation from the torsion, shift of the rotational axis with scan, and possible appearance of spurious elastic modes when lightly damped. Most of these factors can be limited or circumvented by design trade-off and application of appropriate installation procedures [Mon].

Many galvanometer and resonant scanners are equipped with elegant position-sensing devices and control systems, which provide high accuracy and repeatability and render a smoothed and extended bandpass. Several magnetic encoders and capacitive and optical position detection schemes are utilized [Mon,Ayl] for integration into servo control systems that approach accuracies to 1 μrad of positional finesse.

4.5.4 The Fast-Steering Mirror

A relatively new availability of low-inertia vibrational scanner is the two-axis fast-steering mirror [New,Ball,Ber,Swe]. The typical configuration is that of a single flat mirror mounted through a flexure suspension nominally parallel to a flat base, so that the mirror may pivot about a central point near its surface. Four electrodynamic voice-coil type actuators are mounted in perpendicular axis pairs between the edges of the mirror and the base. The actuator pairs are driven in push-pull to nutate the mirror about its quadrature axes, against the restoring force of the flexure suspension, rendering two-axis angular scan.

Commercially available devices exhibit interesting characteristics. Although the scan angle is currently relatively low (to ±3° mechanical), mirror sizes can be substantive (25-mm round to 147-mm square). They are available in materials such as aluminum, beryllium, or glass. As in the closed-loop galvanometer, mirror position sensing with controlled feedback is incorporated to normalize the bandpass and to provide high positional accuracy. Typical characteristics (see, e.g., [Ball] Mdl. 3B) are: beryllium mirror size, 70-mm square; mechanical scan angle, ±1.5° (6° total, optical); resolution finesse, 1-μrad rms; bandwidth, 1000 Hz (3 db); acceleration, 1000 rad/s^2; x-y cross-coupling <0.1%. The overall mirror assembly size is 5 in. by 5 in. by $2\frac{1}{2}$ in. thick, and its mass is 485 g (1.07 lb). Another typical device (see, e.g., [Swe] two-axis FSM) has mirror size, 35-mm round; mechanical scan angle, ±3° (12° total, optical); bandwidth to 1000 Hz; control system, type 2 position servo.

Complementing the limited angular deflection is their significant mirror size. The first sample unit achieves the substantive scanned resolution (Equation 3-5) of 7000 elements/scan in the visible-near-IR wavelength region, assuming for simplicity $a = 1$ and $\lambda = 1$ μm, $\Theta = 0.1$ radian, and $D = 70$ mm uniformly illuminated. The round mirror of the second unit calls for $a = 1.25$ (Table 3.1). Although half the width, it executes twice the scan angle, yielding the substantive resolution of

5600 elements per scan at the same wavelength. Considering their utility and compactness, this technology might be viewed as a contender in the pursuit of "agile beam steering," which is achieved by novel substantively nonmechanical means. Agile beam steering is developed and reviewed comprehensively in Section 4.10.

4.5.5 The Fiber Optic Scanner

A novel method of resonant optical scanning, introduced initially for endoscope application [Sei], was expanded recently [Fau] for more general utilization. It consists simply of a single-mode optical fiber, suspended cantilever for a short distance, and driven at its (bending mode) resonant frequency by a piezoelectric (bimorph) actuator. The other (fixed) end of the optical fiber is coupled appropriately to a light source.

When the fiber is driven as above by a single actuator, the cantilevered output end executes an harmonic scan that is arced from an effective center of curvature, while light propagates outward from the oscillating end. In this one-dimensional mode, it has demonstrated reasonable freedom from perturbing crosscoupling, maintaining high conformity to movement in a single (x-z) plane. When driven by a pair of (x-y) actuators, it forms a proper Lissajous pattern, notably either circular or spiral for information scanning. Both speed and scan angle are high. One-dimensional scan tests have attained speeds above 20 KHz and full scan angles of over 70°, with a displacement of the fiber tip of approximately 1 mm.

Processing of the fiber entails two operations, one more common to all applications and one dependent on its utilization. Starting with a standard optical fiber, the common process entails the narrowing or tapering of its diameter by "pulling," etching, or in general micromachining, as conducted for near-field microscopy [Mur]. This allows the attained high resonant frequencies. The use-dependent process entails the formation and coupling of a microball lens to the output tip of the fiber to converge—to "collimate" and/or to focus the initially widely diffracted output beam for subsequent scanning of an information field.

The ball lens is formed by controlled melting [L&B] of the tip of the cantilevered end of the fiber with a CO_2 laser system. The fiber is mounted in an x-y positioning stage with the tip facing downward, centered along an axis coincident with that of the CO_2 laser facing upward. An axial vibration technique is implemented whereby the fiber end is accelerated along the axis (by the armature of an audio speaker) driven by a sine wave generator. A pulse generator, synchronized with the sine wave generator, triggers the CO_2 laser to pulse at the same point of acceleration of the fiber along the axis. This allows, under control of these time and laser energy parameters, adjustment of the lens melt to vary the size, shape, and focal length of the added lens. With a ball microlens mass added to the oscillatng fiber tip, the resonant frequency is lowered, reducing the 20-KHz (unloaded) scan frequency to one greater than 17 KHZ—still respectable.

When the added ball lens focuses to a sampling point of size δ' (i.e., non-diffraction-limited actual spot size), the device may be considered to function in one dimension as an angular scanning objective lens (arced objective scan per Section 1.5.1.3). In this form, it renders optical scan along a correspondingly arced information surface (such as a miniature drum scanner). When the ball lens propagates a con-trolled beam characteristic into a subsequent flat field-type lens, this is considered preobjective scanning, for the subsequent objective lens forms the final focusing of the spot δ' across the information surface.

Whereas the scanned resolution of this novel deflection technique may be determined experimentally as $N = W/\delta'$, the ratio of the scanned image length W to the spot size δ' at an appropriate spot overlap cri-terion, there is an informative interpretation of the intervening process. As in Section 3.4, this system falls into the class of scanners provid-ing augmented resolution, that is, rendering resolution not only due to the angular deflection of a focused beam N_Θ, but also due to the "trans-lation" of the source, N_s. For diffraction-limited systems, N_Θ is the familiar resolution expressed by Equation 3-5 as $N_\Theta = \Theta D/a\lambda$, which, when augmented with the N_s component, forms Equation 3-10 as $N = (\Theta D/a\lambda) \cdot (1 + r/f)$. Referring to Figure 3.5, the parameters of the system are represented by the fulcrum at point o of the arced scan locus, the distance r to the output surface of the ball lens having a beam width D, its focal distance to P (whether imaging directly with spots δ' on to an arced surface or launched into a subsequent lens with a beam that would focus at P if the lens were not there), and the total scan angle Θ. Proper handling of the focal distance f is expressed at the end of Section 3.4.1, taken as positive for a converging beam and negative for a diverging beam, which is quite possible in this configu-ration when launched from a small D into a subsequent lens. If the beam remains reasonably collimated between the two lenses, f is taken as ∞. As in all calculations of scanned resolution, appropriate correction is instituted for known departure from diffraction-limited performance.

4.6 SCANNER-LENS RELATIONSHIPS

This topic concentrates on those techniques that render large defletion angles to the beam. Thus it involves primarily the rotational scanners, typically the polygon and holographic devices and the vibrational scanners of the galvanometer and resonant types that deflect significant beam angles. A "significant" scan angle into a lens is here considered that which requires serious attention to alleviate off-axis lens aberrations and distortions, often with the use of compound optics. A typical lens configuration dealing with this situation appears in Figure 3.6.

4.6.1 Scanner-Lens Architecture

Objective lens complexity (given wavelength, beamwidth, and scan angle) relates primarily to its orientation before or after the scanner. This is introduced and presented in Section 1.5.1, describing the options of postobjective and preobjective scanning.* It is preobjective scanning that imposes the major burdens on lens design and fabrication. A further parameter is the extent of the pupil relief distance, identified in Figure 1.8 and 1.9. Under similar conditions of beamwidth and scan angle Θ, the greater the pupil relief distance, the larger and more costly is the lens assembly. It requires a larger "clear aperture" for clear propagation of the fully scanned beam through the optical elements.

As detailed in Section 1.5.1, a pyramidal rather than prismataic scanner, also operating preobjective as in Figures 1.9 and 3.6, allows minimization of the pupil relief distance by eliminating the space allocated for the input beam to clear the lens housing. This extra space is evident in Figures 1.8 and 3.7, and Figure 4.2 illustrates the design parameters relating to, for example, Figure 3.7. Keeping in mind that clear passage of a Gaussian beam requires allocating a channel width of almost twice the $1/e^2$ width of the beam (Equation 2-6), the pyramidal configurations of Figures 1.9 and 3.6 can provide reduced lens size and cost.

A similar reduction in pupil relief distance can be achieved with the use of transmissive deflectors (or facets) instead of the above-described reflective ones. Such is the case with transmissive holographic scanners, exemplified by Figures 4.14 and 4.16. The bow compensation prism of Figure 4.16 inhibits an even closer placement of the F-Θ scan lens to the hologon disk.

---

* Although this presentation relates to active scanning (Section 1.2), it applies equally to passive systems (observing the change in related nomenclature), per Figures 1.1 and 1.2, where the conjugate fixed and moving focal points are oriented at the same sides of both systems.

In contrast to the demands of wide-angle preobjective scanning on the design of the objective lens, postobjective scanning imposes no such requirement. This operation is represented fundamentally in Figure 1.10, where the objective lens is not only illustrated as a single element, but may, in reality, be reduced to such a simple component. The aberration burden on the lens is minimized with the input beam propagating coaxially with the lens, avoiding *all* off-axis aberrations. For typical monochromatic operation, the only defect of concern is spherical aberration. Although this could be challenging for focusing from a high numerical aperture (low f-number) to extremely small spots (Equation 2-17), as depicted, for example, in Figures 4.4, even this can be alleviated significantly with the tilted axis alternate of Figures 4.5 and 4.6. These conservation factors are discussed in Section 4.3.5.4.

4.6.2 Double-Pass Architecture

Double-pass operation allows reduction of the pupil relief distance and elimination of some beam expander optics. It functions similarly with over- or underfilled facets. One basic form is introduced in Section 1.2.2.2, shown schematically in Figure 1.4, and illustrated realistically in Figure 4.3. This method is one of two generic forms of double-pass operation. The second form, illustrated in Figure 4.21, is discussed

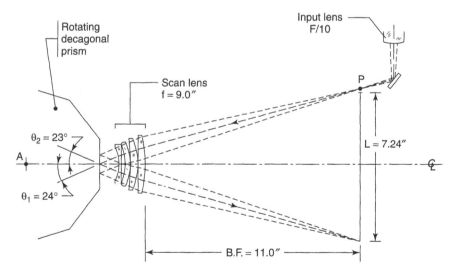

Fig. 4.21 Alternate double-pass configuration—edge input. Compare with Fig. 4.3. Input beam enters from point P, just outside the scan field of length *L*. All beams reside in plane that is oriented perpendicular to the rotating axis A. Numeric values represent a typical design case. From R.E. Hopkins and M.J. Buzowa, "Optics for laser scanning," in *Laser Scanning and Recording, SPIE Milestone Series*, Vol. 378 (1985). Reproduced by permission of the publisher.

below. In the first form, utilizing "central" illumination, the input beam (Figs. 1.4 and 4.3) is directed in the vertical meridional plane of the lens assembly to arrive ideally centered in the scan direction. However, it is displaced *angularly* in the y-z plane (Fig. 1.4), skewed by an amount β/2, which is to be minimized.

There are two principal effects in this first illumination technique. By propagating nominally centered in the *scan* direction of the lens system, the beam arrives at the facet suffering essentially no off-axis aberration *in that direction*. Consequently, the remaining significant concerns are the *cross-scan* aberration, which is compounded by the skewed (fixed) input and scanned output beams, and the magnitude of the residual bow (cross-scan sag) resulting from the skewed traversal of the scanning beam through the lens. Here are challenging tasks for computational exercise to determine and minimize the aberration and bow [H&S].

A useful indication of the functional variations of scan bow is attained by starting with the relation

$$e = f \sin\alpha[1/\cos\Theta_{1/2} - 1] \qquad (4\text{-}24)$$

in which e is the bow displacement (sag) in units of the lens focal length f, α is the deviation angle of the skewed beam with respect to the axis (β/2 in Fig. 1.4), and $\Theta_{1/2}$ is half the full-field scan angle Θ (per Fig. 4.2).

For a rough estimate of the skew angle α, consider that the off-axis focal point in the image plane (p_o or p_1 in Fig. 1.4) is displaced from the axis by an amount equal to the beam height (in the y-direction) at the lens. Figure 1.4 reveals this as a conservative value for the illustrated f-number beam; approximately f/10. This allows easy insertion of a folding mirror per Figure 4.3. It only becomes challenging when the beam height (hence the displacement allotted to the focal point) is reduced to a few millimeters, burdening the shape and fit of the folding mirror.

When considered in this manner, and assigning the beam height H, then the deviation angle α may be represented as $\alpha \simeq \sin^{-1}(H/f)$. This reduces Equation 4-24 to

$$e \simeq H(1/\cos\Theta_{1/2} - 1), \qquad (4\text{-}24a)$$

eliminating the sin α factor and independent of the focal length. Evaluation at the typical beam height $H = 5\,\text{mm}$ and $\Theta_{1/2} = 30°$ (a "large" scan angle into a flat-field lens) reveals a bow sag of $e = 0.775\,\text{mm}$. At the "small" scan angle of ±10° (with the same H), the sag reduces to e

= 0.077 mm—1/10th of that at the ±30° scan angle! And, of course, e varies in proportion to H in Equation 4-24a.

If the residual bow is considered to be excessive, one may compensate this with a complementary bow generated by other means, for example, adding a refractive prism in the beam path between the scanner and lens [H&S,Kra] or altering the facet angles of the (initially prismatic) deflector to provide small pyramidal angles with respect to the rotating axis [Sch,H&S].

The alternate method for forming double pass, noted above in this section, generates no bow. However, it imposes off-axis aberration in the scan direction. Observation of Figure 4.21 reveals that double pass is achieved in this method by injecting the input beam toward the flat–field lens from a point P just outside the nominal scan field (of length L). The system is almost symmetric about the centerline and free of cross-scan error (in y-direction of, e.g., Fig. 3.7). This is valid so long as the input beam and the scanned output beam propagate in the same plane (e.g., the plane of the paper in Fig. 4.21) and be perpendicular to the rotating axis A of the prismatic polygon (or single mirror deflector of Fig. 1.8).

The off-axis aberration in the scan direction is initiated by the input beam (injected at point P) as it propagates through the most marginal portion of the lens. Because this beam is aberrated initially with some astigmatism and coma and remains invariant during subsequent reflection from the deflector, it injects this aberration function throughout the entire output scan process. This initial aberration is then compounded by (and not necessarily complemented by) any new variable off-axis aberration created during scan of the output beam through the lens. Here, again, is an opportunity for computational exercise, likely of comparable effort to that required for the first-described skew beam method.

4.6.3 Aperture Relaying

This topic, seldom addressed in the context of optical scanning, merits attention. Aperture relaying provides operationally efficient transfer of angular scan from one deflector to another. A prominent example is the formation of two-dimensional angular scan by individual x and y deflectors.

Without aperture relay, let the x-deflector execute its required component of scan, which is conveyed over some finite distance to the y-deflector, which provides its component of scan to form the composite function. It is the "some finite distance" that is of concern. Depending on the distance and the scan angle of the x-deflector, the displacement of the beam on the y-deflector may require an enlargement of its aperture to accept the beam. This could affect significantly its dynamics and cost and, perhaps, its realizability. Even if practical from these viewpoints, the resulting deflection from two disparate centers of scan may burden subsequent optical imaging.

Although aperture relaying requires additional optics, it resolves the above concerns. An aperture relay transfers the dimensions and the axis of angular scan of a first deflector to a second location. This is rendered in Figure 1.7 in its most familiar form. Without deflection, it is recognized as beam expansion or compression. Shown as a beam compressor, it reduces the input beam width D to the output width D'. As a relay, it transfers aperture D at focal distance f_1 from lens L_1 to a reduced aperture width D' at reduced distance f_2 beyond lens L_2. The total distance is $2(f_1 + f_2)$. In reverse, it provides beam expansion. Note the complementary transfer maganitudes of D and Θ. For $f_1 = f_2$, $D = D'$ and $\Theta = \Theta'$. Each lens operates telecentrically. As for any substantive angular scan, lens design is to be consistent with acceptable off-axis aberraton.

For unity magnification ($D = D'$), another lens arrangement may be considered (see [Bei2] pp. 129–132), which utilizes a $1:1$ conjugate image lens and a recollimating lens at D'. This was adapted to reflective optical transfer and became a *split confocal resonator*, arranged to provide *scan amplification*, increased angle and resolution through multiple reimaging of a pair of small deflecting mirrors.

4.6.4   Lens Relationships for Control of Deflection Error

Chapter 5 concentrates on this topic, which includes major attention to the utilization of anamorphic (cylindrical) optics for the control of cross-scan error. This subject is influenced, in turn, by the principles of scanned resolution, discussed in its own Chapter 3. Some of these error relationships are introduced in Chapter 3 under Section 3.3.1, Propagation of Noise and Error Components.

4.7   LOW-INERTIA SCANNING

In contrast to the high-inertia devices discussed in all but Sections 4.5.1 and 4.5.4 (the galvanometer scanner and the fast-steering mirror), those considered further are cast firmly in the low-inertia domain. Recalling the anomaly of the resonant scanner having a low-inertia armature but operating functionally as a high-inertia device (Section 4.5.2) and so charted in Figure 4.1, the following techniques are truly low inertia in performance. They may be classed as *"agile,"* capable of rendering rapid alteration of position, velocity, or direction to the scanned beam. Along with the more familiar acoustooptic (AO) and electrooptic (EO) devices (including prism gradient and digital binary) are techniques accorded more recent attention, identified in Figure 4.1 as microlens array and phased array.

4.8 ACOUSTOOPTIC SCANNERS

This device, utilizing the acoustooptic effect (sound wave induction of refractive index variation in a medium), is one of the most diverse of low-inertia scanners. With appropriate material selection, it serves a wide range of optical wavelengths. Transmission ranges of several materials overlap to cover substantial portions of the ultraviolet-visible-infrared spectrum; exemplified by lithium niobate ($LiNbO_3$, 0.04–4.5 μm), telurium dioxide (TeO_2, 0.35–5 μm), and germanium (Ge, 2–20 μm). Although acoustooptic devices are also designed for and suited ideally to intensity modulation of laser beams, this work concentrates on its scanning characteristics. Before the advent of economical and well-controlled modulatable laser diodes, almost every "laser printer" utilized an acoustooptic modulator. It is noteworthy, too, that almost no laser printer utilized acoustooptic deflection. This illustrates one of the trade-offs of low-inertia devices: limited capability for forming adequate resolution along a contiguous scan line of typical page width. [It may be recognized that a similar limitation exists for the broadband galvanometer (Section 4.5.1), while scanning at a sufficiently high rate.] If the limitation of scan line continuity is relaxed, one may compose and join narrower subraster columns to form the full width. This technique (called *stitching*) is, however, seldom implemented because of the exacting requirements for assembling the columns to such accuracy in position and image density that their joints are practically imperceptible under loupe scrutiny.

4.8.1 Operating Principles

An acoustooptic deflector showing typical beam propagation is illustrated in Figure 4.22. Although several interactions of light and sound have been investigated intensively [Adl,D&Se,Kor,Gor,Bed], these devices operate in the Bragg diffraction domain, in a manner described

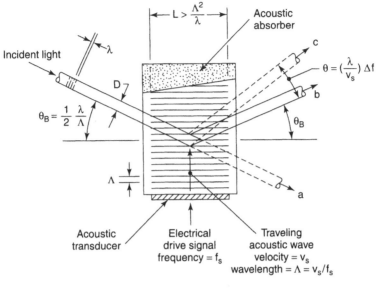

Fig. 4.22   Acoustooptic deflector. Angles exaggerated for illustration. Electrical drive upon acoustic transducer generates traveling acoustic wave in elastic medium. Resulting periodic index change simulates a thick optical grating. Input beam and output beam (at position b) form nominal Bragg angles. Relationship between the output beam positions and the acoustic frequency is tabulated. From [Bei8].

initially in Section 4.4.1 for some disktype holographic scanners. Although there is a distinction in the type and thickness of grating (a surface relief grating on the holographic disk and a volume grating of substantive thickness in the AO device), the diffraction properties are similar.

As rendered in Figure 4.22, a diffraction grating of spacing Λ is synthesized by the wavefront spacings of an acoustic wave traveling through a block of elastic material (which is transparent to the optical wavelength of interest). At the base of the block is fastened an acoustic (piezoelectric) transducer that converts an electrical drive signal to a pressure wave that traverses the medium at its acoustic velocity v_s. At the far end, the energy in the traveling wave is consumed by the acoustic absorber to suppress standing waves. The pressure wave in the medium forms a corresponding photoelastic variation in its refractive index, forming the synthetic optical grating. An incident light beam of width D is introduced at the Bragg angle Θ_B (shown exaggerated). Although the above-developed index variation is extremely small, the cumulative effect of the beam propagating through the long path within the medium (thick grating of length $L > \Lambda^2/\lambda$) provides for a substantive diffraction efficiency. In the perfect Bragg condition, the energy in all pairs of diffraction orders is transferred to one first order at the output Bragg angle Θ_B, attaining a theoretical diffraction efficiency of 100%. Operationally, depending on compliance of Bragg, beam shape and polarization, and optical wavefront purity, drive signal characteristics, and material and coating losses, the diffraction efficiency can range between 50% and 90%.

In Figure 4.22, the Input Beam is incident at the Bragg angle, $\Theta_B = \frac{1}{2}\lambda/\Lambda$. The undeflected output beam at (a) is the undiffracted *zero order*. The output beam at (b) is the diffracted *first order*, at the Bragg angle Θ_B when the drive signal is at center frequency f_o. The dashed-line ouput beam at (c) is the deflected beam displaced "positively" from position (b) when f_o is increased to $f_o + \Delta f$, thereby decreasing the grating spacing. The "negatively" deflected beam (not shown) appears on the other side of (b) when f_o is decreased to $f_o - \Delta f$, increasing the grating spacing. [Note: Small delta (Δ) modifying (f) signifies the *change* in frequency; large delta (Δf) represents the full frequency bandwidth.] The full Δf is typically limited to one octave to avoid formation of "ghost" diffraction spots that can be generated by harmonics of the acoustic frequency.

If the Bragg angle is set symmetrically at center frequency f_o, then when Δf is instituted the output beam is deflected, altering and disrupting the perfect Bragg condition. Unless rectified as indicated below, this depletes the transfer efficiency, which can be reduced by about 1/3 of its initial values as the $|\Delta f|$ is shifted to its typical limits of about $\pm\frac{1}{4}$ of its center frequency. Although seldom applied, a direct attack on the source of this problem is altering the direction of the acoustic wave to track and maintain bisecting the incident and diffracted beams during scan [Kor,CG&A,Got]. A principal method utilizes a phased array of acoustic transducers, in a manner discussed for optical scanning in Section 4.10.1. Acoustic beam steering for Bragg angle sustenance was implemented successfully as early as 1966 by A. Korpel and his associates [Kor] at Zenith Radio Corporation (and witnessed in their laboratory by this author), in their pioneering develpment of an acoustooptically deflected (15,750 scans/s) projection laser display system. The acoustic medium was water.

4.8.2 Fundamental Characteristics

As with the grating equation applied in holographic deflection (Eq. 4-22), diffraction from a structure having a periodic spacing Λ (Fig. 4.22) is expressed as

$$\sin\Theta_i + \sin\Theta_o = n\lambda/\Lambda \qquad (4\text{-}25)$$

in which Θ_i and Θ_o are the input and output beam angles, respectively, n is the diffractive order, and λ is the wavelength. As a "thick" diffractor of length

$$L > \Lambda^2/\lambda \qquad (4\text{-}26)$$

all the diffracted orders are transferred efficiently to one first order. Because Bragg operation requires that

$$\Theta_i = \Theta_o \equiv \Theta_B \qquad (4\text{-}27)$$

then from Equation 4-25, for the typically small angles of acoustooptic deflection, the Bragg angle reduces to

$$\Theta_B = 1/2\lambda/\Lambda \qquad (4\text{-}28)$$

and the full scan angle becomes

$$\Delta\Theta = \lambda/\Delta\Lambda = (\lambda/v_s)\Delta f \qquad (4\text{-}29)$$

The beamwidth D is traversed by the acoustic wave during its transit time τ at the acoustic velocity v_s such that

$$D = v_s\tau \qquad (4\text{-}30)$$

The transit time $\tau = D/v_s$ also represents the access time, or the time required to institute a change in beam direction for random access applications.

Using the relationships for scanned resolution and duty cycle (Equations 3-5 and 4-1, respectively) and substituting Equations 4-29 and 4-30 for Θ and D, the resolution of the acoustooptic deflector becomes

$$N = \frac{\tau \Delta f}{a}\left(1 - \frac{\tau}{T}\right) \tag{4-31}$$

The $\tau \Delta f$ component represents the time-bandwidth product, a measure of information handling capacity and analogous to the ΘD product in Equations 1-3, 1-4, and 1-6. Insertion of typical parameters into Equation 4-31 leads to a resolution limit of approximately 1000 elements per scan. Note that a wide beamwidth and a low acoustic velocity are required to attain a large τ for high resolution. Yet a short access time demands the inverse conditions. This also illustrates the conflict between AO deflection and AO modulation, each requiring inverse conditions in the τ for high deflector resolution and high modulation rate, and that AO deflection requires acoustic frquency modulation (FM) and AO modulation entails acoustic amplitude modulation (AM). A brief historical review, a variation on the treatment of scanned resolution, and discussions of acoustic power effect on diffraction efficiency, the application of anamorphic optics to reduce the power requirement, and the selection of AO materials, were presented in an earlier work [Bei2].

4.8.3 Alternate Acoustooptic Deflection Techniques

4.8.3.1 The Scophony Scanner Although AO deflection is dominated by the implementations discussed above, several variations are significant. The earliest investigated, only five years after the theoretical exposition of acoustooptics in 1932 [D&Se], was the technique applied to the Scophony TV receiver [Oko]. This more expansive concept, which remains known as the Scophony process, was analyzed more completely [Joh,JG&S] and continues to be encountered in varying amounts of application of its principles.

All active scanners considered here—other than the Scophony—are known as flying spot scanners, as introduced in Section 1.2. That is, they all cause the motion of a single spot or image element, typically "flying" at a high velocity. This descriptor has been used since the early years of cathode-ray tube (CRT) scanning [B&Y]. In the Scophony scanner, however, the acoustic traveling wave usually encompasses more than one image element. Although the earliest Scophony scanner was illuminated with incoherent light, and the imaged multiple spots remained incoherently related, one must now consider the consequence of illuminating the AO device with a laser, spatially coherent.

When the laser-illuminated AO device scans a single spot, although the beam is spatially coherent, its motion in the image plane destroys the relative coherence of adjacent spots, resulting in spot resolutions typified by an *incoherent* source. The resulting modulation transfer functions (MTFs) also remain as of an incoherent source, and are also so represented in Chapter 3. However, the virtue of the Scophony scanner depends on the acoustic wave operating on a broad illuminating beam subtending several information elements. This is accomplished as the acoustic wave, now modulated by an information stream, propagates through this broad beam, affecting the medium correspondingly in elements of grating intensity. The modulating acoustic drive signal is now characterized as AM, not FM. The Scophony process is a *modulation*, which causes a stream of diffractive elements to travel at the acoustic velocity. When this pattern is reimaged, it will travel at this velocity magnified by any optical magnification. Thus the Scophony modulator requires a subsequent deflector to cancel the motion of the rapidly moving array of spots to form a stationary image. Consider Figure 4.23.

A laser and beam expander illuminate the Scophony modulator with, as indicated above, a broad beam. Otherwise, the configuration is very similar to that of the AO deflector, while the acoustic drive center

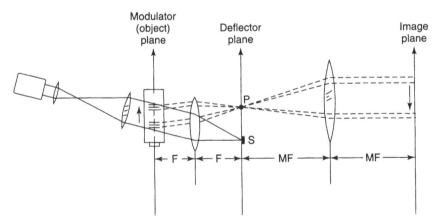

Fig. 4.23 Scophony scanner optics. Expanded laser illumination enters modulator as in Fig. 4-22. Two bursts of acoustic wave, traveling "upward," cause diffraction of two beams that are then focused by lens of focal length F at point P on deflector plane. Underviated zero order is also focused at deflector plane, where it is absorbed by stop S. Diffracted beams continue through 2nd focal distance MF (M = magnfication) twice, arriving at image plane traveling "downward." Deflector (not shown) in vicinity of point P scans beams in reverse direction to immobilize them on the image plane. Subsequent scans lay down subsequent beam positions in a raster, forming a correct image. From [Joh].

frequency (f_o in Fig. 4-22) remains constant. When not driven (f_o off), the zero order propagates through directly. When driven, the diffracted components exit, satisfying the nominal Bragg condition for high efficiency. The illustration represents a binary on-off acoustic drive, simulating 100% AM modulation. Two bursts of drive, forming two diffraction gratings, are shown traveling upward through the medium at the acoustic velocity. The exiting diffracted beams and the zero-order beam encounter a first lens of focal length F such that they focus at the surface marked "deflector plane." Because both diffracted beams are collimated and are incident at the lens at the same angle, they form a single focal point P. The zero order is blocked by a "stop." The two diffracted beams continue toward a second lens of focal length MF (M = optical magnification) to image two related moving spots on the image plane. A "conventional" deflector is now added in the vicinity of point P to scan the incident beams in a reverse direction at such speed as to immobilize the output image. Thus a stationary object that is represented by the previously scanned moving data points in the Scophony modulator will be portrayed properly as a stationary image. The inverse velocity must be exact; otherwise, cancellation will be imperfect, and the image will drift.

This imaging system is a simple form of schlieren optics, in which the scattered (diffracted) information components are rendered clear optical passage. The focal region near P is a good location for a deflector because of its small size (exhibiting the Fourier transform of the grating transmission). Derived from a large diffraction region, it converges and expands rapidly around focus (depending on f-number), motivating use of a mirrored deflector such as a polygon scanner, rather than one having substantive thickness, such as an AO scanner. The relative coherence of adjacent spots, identified above, now merits attention. Fully coherent imaging of high-contrast objects (such as bar targets) yields significant edge ringing, degrading resolution [P&T,Joh]. This may be alleviated to the degree that coherence is reduced. Some Scophony systems reduce the number of data elements appearing within the diffracting aperture, thereby approaching the resolution characteristics of coherent single spot scanning noted above. An objective of the Scophony system is to maintain a wide illuminated aperture to maximize acoustooptic modulation efficiency.

4.8.3.2 The Traveling Lens and Chirp Deflectors

Both the traveling lens and the chirp deflectors utilize acoustooptics in a functionally similar manner that differs significantly from the conventional AO deflector discussed in Sections 4.8.1 and 4.8.2. Instead of diffracting an incident beam through an angle that is determined by the drive frequency, a pressure wave, developed in the medium by a *fixed frequency pattern*, shapes the index variation in the medium to appear as a small lens that travels through the medium at the acoustic velocity. Illumination of this traveling simulated lens with a narrower beam forms a smaller focal point that travels across the full aperture length, forming a long linear scan at the high acoustic velocity and providing notably high resolution. A conventional AO deflector prescans the beam that illuminates and tracks the traveling synthetic lens in the linear scanner. Allowing significantly higher resolutions, these systems still serve as flying spot scanners.

The traveling lens concept was described in 1970 [FC&C] by Zenith Radio Corporation, followed by intensive investigation by the Harris Corporation [J&M]. The basic process is illustrated in Figure 4.24,

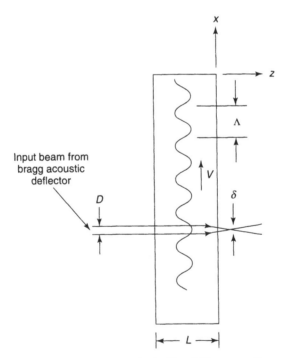

Fig. 4.24 Basic traveling lens process. As in Fig. 4.22, a traveling wave propagates "upward" at velocity V while an input beam of height D is scanned at the same velocity to track a crest in the harmonic pressure wave. The resulting focused spot of size δ $\ll D$ scans also at velocity V, multiplying resolution by the factor D/δ. After R.J. Johnson and R.M. Montgomery, "Optical beam deflection using acoustic-traveling-wave technology," in *Acousto-Optics/Instrumentation/Applications, Proc. SPIE*, Vol. 90 (1976). Reproduced by permission of the publisher.

where a cell of acoustooptic material of thickness L exhibits a traveling sinusoidal pressure wave propagating at velocity V along the X-direction (generated by a piezoelectric transducer driven by a sinusoidal r-f frequency that establishes wavelength Λ within the material.) An input beam of width D enters the cell at a location coincident with the crest of one harmonic presure wave and tracks this crest at velocity V as described above. When the width of the beam satisfies $D \simeq \Lambda/4$, the beam encounters a near-parabolic index change, acting as a refractive cylindrical lens of length L. This provides for acceptably low-aberration focusing of the beam to the line width δ such that $\delta \ll D$. Thus the resolution developed originally at the AO deflector is multiplied by the factor D/δ in the focused X-direction. Additional cylindrical focusing in the quadrature y-direction is utilized to reshape the "line" spot to round or elliptical, as required. Resolution gains of 40 have been achieved over a scan length of up to 10 in. [J&M].

Figure 4.25 illustrates the scanning and optical transfer arrangement of a traveling lens system. A broad illuminating beam enters a conventional AO beam deflector at the left. In the upper illustration, the angular change at the output corresponds to $\pm\Theta$ in Figure 4.22, with the AO aperture fully illuminated. Bragg angles at the input and output are omitted for clarity. Scan direction is such that the solid-line beam moves toward the dashed-line beam. Lens L_1 is oriented approximately telecentrically ($Z \simeq f_1$) between the AO deflector and its traveling focal point, transforming the angular scan to a linear scan. This scanning focal point in plane P_1 is then relayed by lenses L_2 and L_3 into the traveling lens cell, plane P_3. There, the synthetic lens (shown as a real lens) starts at the bottom of the cell illuminated with the beam height D as in Figure 4.24, which tracks the lens as it travels upward at velocity v_s over the magnified scan distance L_s. Lenses L_1 and L_2 may be regarded as a beam compressor, acting in a manner very similar to that in Figure 1.7, thereby magnifying the small scan angle of the AO deflector (shown in Fig. 1.7 as $\Theta/2$ enlarged to the output angle $\Theta'/2$, as noted in the caption). At Plane 3, the traveling lens focuses the incoming collimated beam to the scanning focal point δ outside the cell, as represented in Figure 4.24. Inside the cell, the traveling wave extends over the full thickness of the cell (L in Fig. 4.24) focusing in a continuum, as by a graded-index cylinder of length L. Because the traveling lens cell provides optical power anamorphically in the plane of the paper, quadrature optical power is provided in and out of the paper by cylindrical lens L_4 to form a corrected two-dimensional spot. The bottom illustration shows the optical action in that plane.

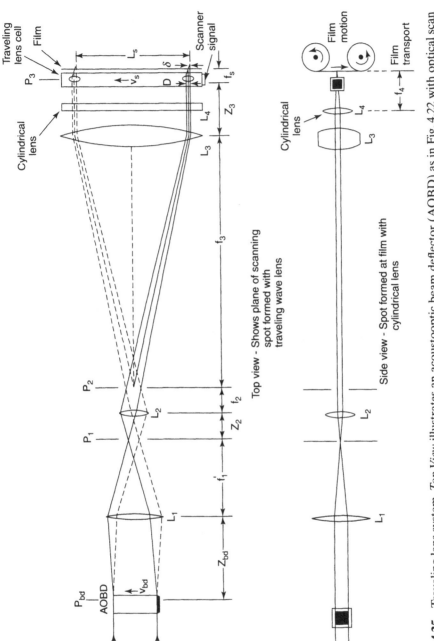

Fig. 4.25 Traveling lens system. *Top View* illustrates an acoustooptic beam deflector (AOBD) as in Fig. 4.22 with optical scan angle magnification as in Fig. 1.7. They provide the prescanned beam of height D that tracks the traveling lens in plane P_3 to focus (per Fig. 4.24) in the plane of the paper. *Side View* shows the contribution of the cylindrical lens L_4, which focuses the beam in the quadrature direction. After R.J. Johnson and R.M. Montgomery, "Optical beam deflection using acoustic-traveling-wave technology," in *Acousto-Optics/Instrumentation/Applications*, *Proc. SPIE*, Vol. 90 (1976). Reproduced by permission

In 1979, workers at Rockwell International and Harris Corporation [YW&M] reported an advance that significantly reduces the substantive electrical power consumed by the traveling lens. This is accomplished by replacing the continuous r-f signal (forming the continuous sinusoid in Fig. 4.24), with a narrow r-f pulse, utilizing its central lobe as an isolated lens traveling at velocity v_s. This conserves the energy otherwise dissipated in the unused portion of the full field of Figure 4.24. Power reductions by a factor of 25 have been observed. Some demanding special requirements must be accomodated, as discussed in the referenced work.

The chirp deflector [Bade] utilized a different r-f drive technique in the traveling lens system, forming the lens diffractively rather than refractively. By driving the scanner with bursts of r-f signal that shape the synthetic lens, it not only conserves power as described above but allows for adjusting the focal length of the lens while maintaining a large acoustooptic index change.

The basis of forming the diffractive lens is gleaned from the grating equation (Eq. 4-25) for small angles, with $\Theta_1 = 0$ and $n = 1$,

$$\sin\Theta = \Theta = \lambda/\Lambda = \lambda f/v_s \qquad (4\text{-}33)$$

indicating that Θ is inversely proportional to the acoustic wavelength Λ or directly proportional to the drive frequency f. When the frequency is altered rapidly such that portions of the incident beam are exposed to progressively varying wavelengths (grating spacings), the beam will converge or diverge, depending on the direction of the frequency change. The lensing is converging for increasing frequency (df/dt > 1), and conversely for decreasing frequency. With a linear frequency change, identified as a "chirp," the focal length of the beam is given by [G&M]

$$f = \frac{v_s^2}{\lambda(df/dt)} \qquad (4\text{-}34)$$

The chirp deflector is illustrated in Figure 4.26a, in which the scanning mode is directly analogous to the operation of the traveling lens cell at the far right of Figure 4.25, having the refractive lens replaced by a diffractive one. With the chirp pulse traveling "upward" at velocity v_s, it is comprised of a decreasing acoustic wavelength, that is, driven by an increasing signal frequency to form a positive lens. The input laser beam scans upward at the same velocity to track the chirp diffractor in its

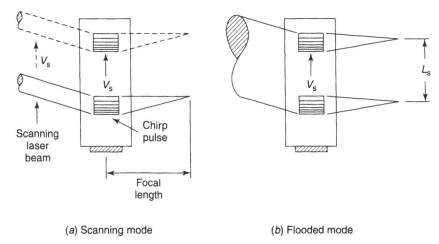

(a) Scanning mode   (b) Flooded mode

Fig. 4.26 The chirp deflector. Utilizes a diffractive traveling lens. The drive signal to the acoustic transducer (Fig. 4.22) forms a positive "chirp"—a rapid increase in frequency—that synthesizes a lens of positive focal length traveling "upward" at velocity V_s. (a) *Scanning Mode*: Prescanned beam tracks the chirp at the same velocity. (b) *Flooded Mode*: Wide beam fills the full aperture, eliminating need for prescan, but sacrificing beam energy while illuminating the nonlens region. After L. Bademian, "Acousto-optic laser scanning," in *Optical Engineering*, Vol. 20 (1981). Reproduced by permission of the publisher.

formation of the scanning focal point. Figure 4.26b illustrates a flooded mode of operation, scanning the same distance L_s. By filling the acoustic aperture with the input beam, as in Figure 4.23, the need for prescanning is eliminated and flyback time may be reduced effectively to zero. As in most overfilled systems except, for example, in the Scophony, the main trade-off is the waste of input beam energy when not illuminating an active lens region.

4.9   ELECTROOPTIC (GRADIENT) SCANNERS

A general form of low-inertia scanner is the gradient deflector [Bei2,Bei6], capable of extremely rapid beam deflection and flyback. Utilizing a controllable gradient of index of refraction across the beam, a propagating wavefront undergoes increasing retardation transverse to the beam. The rays, as the orthogonal trajectories of the wavefronts, bend in the direction of the shorter wavelength. Referring to Figure 4.27a, the small bend angle Θ through such a deflection cell is expressed as

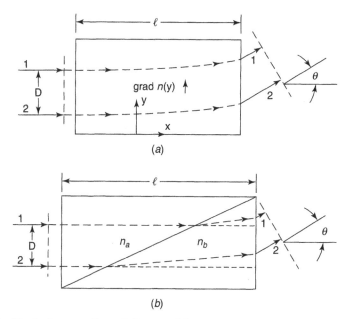

Fig. 4.27 Equivalent gradient deflectors. [*a*] Basic deflector cell having continuous optical index gradient, grad $n(y)$. Ray 1 propagates through a higher index (effecting a greater retardation) than ray 2, rotating the wave about an effective z-line pivot located at the approximate center of the cell. [*b*] Analogous prismatic cell, in which $n_a > n_p$ such that ray 1 is retarded more than ray 2, tipping the wave through the deflected angle. The deflected angles are enlarged slightly by the index change at the output. From [Bei8].

$$\Theta = k(dn/dy)\ell \tag{4-35}$$

where n is the number of wavelengths per unit axial length ℓ, y is the transverse distance and k is a cell system constant. This relationship may be generalized to the gradient deflector equation with vector notation [Bei6]

$$d\Theta = k \text{ grad } n \times d\ell \tag{4-36}$$

in which

$$d\Theta = d\Theta \text{ } \mathbf{z} \quad (\mathbf{z}\perp\mathbf{y},\mathbf{x}) \left.\right\}$$
$$d\ell = d\ell \text{ } \ell \quad (\ell\rightarrow\mathbf{x}) \left.\right\} \text{(see Fig. 4.27}a)$$
$$\text{grad } n = (n/y)\mathbf{y} + (n/x)\mathbf{x} + (n/z)\mathbf{z}$$

For the electrooptic material form, in which the wavefront traverses a gradient of index of refraction over the differential distance $d\ell$, the differential in deflection angle within the material becomes

$$d\Theta = (1/n_o) \operatorname{grad} n \times d\ell \tag{4-37}$$

where n is the index of refraction, n_o is its average value, and grad n is the rate of change of n in the direction of deflection. It is noteworthy that n now represents the index of refraction, whereas above, n represented the wavelength density. The full scan angle may now be expressed for the condition of the light waves traversing index change Δn over the beam aperture D in a cell of length L, to develop the typically small deflection angle

$$\Theta = (\Delta n / n_f) L / D \tag{4-38}$$

where n_f is the refractive index of the final medium (forming, per Fig. 4-27, the refractive angular change of the beam as it leaves the cell). Utilizing our basic Equation 3-5, the corresonding resolution in elements per scan for a given Δn is now shown independent of D as

$$N = (\Delta n / n_f) L / a\lambda \tag{4-39}$$
$$(= L\Delta n / a\lambda \quad \text{for} \quad n_f = 1)$$

The Δn, determined by the material and its applied field, is given by

$$\text{for class I materials} \quad \Delta n = n_o^3 r_{ij} E_z \tag{4-40a}$$
$$\text{for class II materials} \quad \Delta n = n_e^3 r_{ij} E_z \tag{4-40b}$$

where $n_{o,e}$ is the (ordinary, extraordinary) index of refraction, r_{ij} is the electrooptic coefficient, and $E_z = V/Z$ is the electric field in the z-direction, perpendicular to the x- and y-directions.

4.9.1 Implementation Methods

A time-dependent index gradient was proposed [G&B] utilizing resonant acoustic pressure variations in a cell of a transparent material. Although this appears similar to acoustooptic deflection (Section 4.8), it differs fundamentally in that the cell is terminated reflectively to support a standing wave rather than absorptively to maintain the traveling wave. The acoustic wavelength is much longer than the beamwidth to impart the refractive effect rather than much shorter for diffraction. An approach to a linearly varying index gradient utilizes a quadrapolar array of electrodes bounding an electrooptic material [Fow,Los].

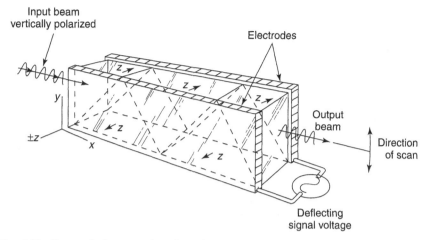

Fig. 4.28 Iterated electrooptic prism deflector. Alternating crystallographic z-axes provide alternating index changes that are varied in magnitude and direction by the applied signal voltage to form the tandem deflecting transitions. A uniform electric field is established in the z-direction by the parallel plate electrodes. [Bei2]

A continuous index gradient can be simulated with the use of alternating electrooptic prisms. A single-stage biprism is illustrated in Figure 4.27b and an iterated array increasing length and effect appears in Figure 4.28. Each interface imparts cumulative retardation across the beam. The direction and rate of retardation are controlled by the electrooptic index changes in the material. Because most electrooptic coefficients are extremely low, high drive voltages are required to achieve even moderate resolutions (to $N \simeq 100$). These devices can, however, scan at very high speeds (to 10^5 per s) while imposing very low time delays ($<0.1\,\mu$, in contrast to the significant τ of AO devices), allowing broadband negative feedback with stable angular control of the beam. Significant review, experiment, and test are reported for the electrooptic deflector [L&Z, Bei2, I&L].

Whereas Figure 4.27 illustrates the passage of a generally collimated beam through the EO material, it is often desired to propagate a convergent beam, to form the focal point just beyond the deflector. There is also motivation, when the beam is proximate to the cell walls and executing significant deflection, to converge the beam inside a long cell to avoid encountering the walls during scan. Anticipating aberration resulting from gradient deflection of a convergent beam, this condition was analyzed and reported [Bei6] to exhibit a direct relationship to the scan angle magnitude and the f-number of the focusing bundle. The results of this work are summarized and analyzed further [Bei2] to include subsequent related consideration of beam convergence in EO deflectors [F&S, L&Z].

4.9.2 Drive Power

The power dissipated within the electrooptic material is given by

$$P = \frac{1}{4}\pi V^2 C f/Q \qquad (4\text{-}41)$$

where V is the applied (p-p sinusoidal) voltage in volts, C is the deflector capacitance in farads, f is the drive frequency in Hz, and Q is the material Q-factor (Q = 1/loss tangent).

The capacitance for the transverse electroded deflectors of Figure 4.28 is approximately that for a rectangular parallel-plate capacitor

$$C = 0.09\kappa LY/Z \quad (\text{pF}) \qquad (4\text{-}42)$$

where κ is the dielectric constant of the material of length L, width Y, and thickness Z.

The characteristics of some electrooptic materials [T&K, Bei2, I&L] are often a strong function of frequency beyond 10^5 Hz. A resolution-speed-power figure of merit has been proposed [Z&L].

4.10 AGILE BEAM STEERING

This chapter has been dedicated, thus far, to technlogy that is familiar operationally to many readers and that has enjoyed broad representation in the field through product optimization for useful application. Looking back to the Figure 4.1 classification of optical scanning technology at the introduction to this chapter, the disciplines that now merit attention are those that complete the bottom row of the low inertia category. The two on the left are represented in a class of growing work called *agile beam steering* [Wat1, McM, W&M, Farn]. Although these techniques were developed initially for such environmentally challenging tasks as laser radar and forward-looking IR (FLIR), continued advancement may allow their cost-effective extension to productive commercial application.

The principal motivation for these developments is to acquire the performance of the angularly articulated mirror while avoiding some of the concomitant burdens of size, weight, inertia, or power supply requirement. This has been a long-envisioned goal of many earlier researchers, in work having important similarities [Bei2] to the more current activity in agile beam steering. Recent research has, quite astutely, harvested and advanced new resources such as liquid crystal E-O phase shifters and micromachining techniques, assembled in novel configurations that embody problem-solving insight. Our objective is to offer a unified expression of this diverse R&D activity. Although many approaches to agile beam steering have been and will continue to be explored, primary attention is devoted to two principal techniques that have dominated investigation and development, the phased array and the decentered microlens array, along with some of their principal variations. Although the basic operation of these arrays differ, the commonality that exists in the formation of the steered wavefronts is discussed.

4.10.1 Phased Array Technology

The broad topic of phased array beam steering merits significant attention because of its operational potential in many options of implementation, extending to some not-so-apparent variations. The positioning of radiation in the radio and microwave regions of the electromagnetic spectrum by driving antenna arrays with controlled relative phasing is especially familiar to the radar specialist [C&F,Sko]. The adaptation of this technology to the optical spectrum, notably as radiated by lasers, was investigated in 1964 [Pro], soon after the invention of the laser, and continued to merit attention [Alp,Gra,Mey] through the early 1970s. The prospect of altering the direction of a laser beam with small adjustments on a group of radiators appeared very attractive. Beam steering with the actuation of arrays of mirrors was investigated by a research team in 1967 [Alp]. With the introduction of electrostatically actuated membrane mirror arrays in 1971 [Gra], and the programming of electrooptic crystal arrays in 1972 [Mey], operational utility was represented. Further attention was directed toward mirror array beam steering in the infrared region, where mirror reflectance exceeds the transmittance of even the exotic infrared materials, and where the longer wavelength imposes less burden on the sustenance of mirror flatness. With the availability of faster-acting materials and novel design variations, substantive advances have been achieved.

4.10.1.1 *Basis of Phased Array Beam Steering*   The alteration of an incident optical wavefront by phase variation is characterized by

considering the effect of a refraction prism. Figure 4.29*a* illustrates a plane wavefront in air, incident parallel to the plane surface of a dielectric wedge. Within the material of refractive index >1, the wavelength is compressed proportionately while the fronts remain parallel to the incident wavefront. A linearly increasing local phase delay results from the progressive retardation of the wavefronts across the enlarging wedge thickness. Upon encountering the tilted boundary, the wavelengths are reexpanded and their angle of propagation is altered as illustrated. This familiar refractive process is exemplified in the prismatic electrooptic gradient deflector of Figure 4.27*b*.

To provide a wide aperture when composed of material having a moderate refractive index, this configuration can exhibit substantive bulk. To relieve this burden, the long wedge profile is divided into an array of smaller wedge periods, each forming a linear phase delay of from 0 to 2π. As illustrated in Figure 4.29*b*, when implemented as described below, this technique provides the same optical deflection as the continuous single wedge. Along with the need to accommodate the combination of slope and refractive index of the material, one must dimension the periods of the wedges such that they form 2π phase differentials (or multiples thereof, i.e., modulo 2π) at the operating wavelength, to assemble continuous nonstaircased wavefronts in the near field. This is functionally analogous to the refractive or the more familiar reflective blazed grating, in which high efficiency is achieved when the angle of specular reflection (from the sawtooth slopes of the grating surfaces) coincides with the angle of diffraction *at the selected wavelength*. Accordingly, an array providing (single or multiple) 2π phase differentials normally exhibits a throughput efficiency having a wavelength sensitivity, thus limiting one composed of fixed full-phase segments to near-monochromatic operation at the selected diffraction order.

The above examples provide continuous phase retardation by virtue of their linear surface slopes, either continuous or in increments. Incremental phase retardation can be achieved by other means, in transmission by an array of small electrooptically controlled cells and in reflection by precise actuation of individual mirrored pistons. A functional illustration of an array of refractive phase retarders and the formation of the radiated wavelets into contiguous wavefronts is provided in Figure 4.29*c*. Operation is similar with pistons, except that the reflective piston requires displacement of only $\frac{1}{2}$ of the 2π phase retardation distance. A thin electrooptic retarder having a reflective ground plane not only requires $\frac{1}{2}$-thickness but, if composed of a relatively slow liquid crystal, attains a fourfold increase in switching speed. This figure illus-

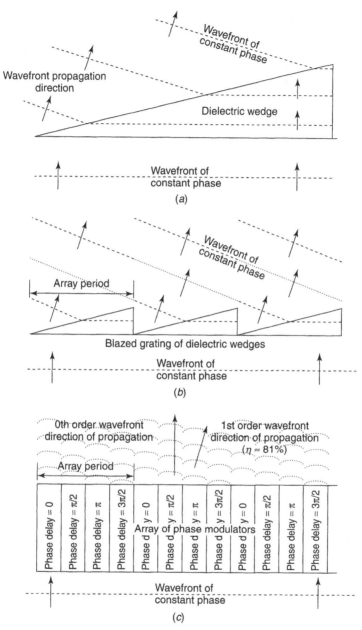

Fig. 4.29 Evolution of optical phased array beam steering in transmission; analogous in reflection. [*a*] Prototype prismatic wedge illustrating familiar refraction of incident plane (constant phase) wavefront. [*b*] Synthesis of [*a*] with array of wedges; each width imparts 2π phase delay over the array period. [*c*] Synthesis of [*b*] with multiple delay elements (4 per 2π period). Superposition of output wavelets forms wavefronts at idealized efficiency of 81%. Greater multiplicity provides higher efficiency. After A.S. Keys, R.L. Fork, T.R. Nelson, Jr., and J.P. Loehr, "Resonant transmissive modulator construction for use in beam steering array," in *Optical Scanning: Design and Application, SPIE*, Vol. 3787 (1995). Reproduced by permission of the publisher.

trates another option: division of the full 2π phase change into a number of substeps, four steps per full cycle in this case, providing 81% diffracted into the first order. The more steps, the higher the efficiency. Eight steps per cycle attains an otherwise lossless efficiency of 95%. Figure 4.29 is a pedagogic unification of three illustrations of [Keys], providing an introduction to the technology. Informative reviews appear in [McM, Mc&W] and in the Air Force report [Dor], Chapter 2.

Some of the heuristic observations expressed above are affirmed by the considerations that follow. The angular relationships of the phased array are expressed by the diffraction grating equation, presented earlier as Equation 4-25 for beam deflection of the acoustooptic grating and represented here for this application,

$$\sin\Theta_i + \sin\Theta_o = n\lambda/\Lambda \qquad (4\text{-}43)$$

where Θ_i and Θ_o are the input and output beam angles with respect to the grating normal (boresight), n is the diffraction order, λ is the free space wavelength, and Λ is the grating (array) period, per Figure 4.29, b or c. As in Figure 4.29c showing four delay elements per array period, for q delay elements, each separated by a fixed distance d, $\Lambda = qd$. Because the number of elements in each period is $q = 2\pi/\phi$, where ϕ is the phase shift between elements, then $\Lambda = (2\pi/\phi)d$, the distance required to assemble a one-wave phase difference. When, as is typical, $\Theta_i = 0$ and we seek the angle of first-order wave propagation (per Fig. 4.29, b or c), then

$$\sin\Theta_o = \lambda/\Lambda \qquad (4\text{-}43a)$$

$$\equiv \lambda/qd = \lambda\phi/2\pi d \qquad (4\text{-}43b)$$

The normalized intensity of the radiation pattern follows the analogy of the one-dimensional microwave phased array [Sko], expressed compactly as

$$L = (\sin N\alpha/N\sin\alpha)^2 \qquad (4\text{-}44)$$

in which

$$\alpha = \pi d/\lambda(\sin\Theta - \sin\Theta_o) \qquad (4\text{-}45)$$

where Θ is the angle with respect to the grating normal at which the field in free space wavelength λ is measured, N is the number of phase shifters in the array, and, as stated above, the uniform elemental spacings d are assumed to provide uniform phase difference ϕ between elements.

A direct analog to Equation 2-2b is that for the efficiency η_q of a linear array having the nominal (blazed) 2π phase resets illustrated in Figure 4.29, b and c; expressed by

$$\eta_q = [\sin(\pi/q)/(\pi/q)]^2 = \mathrm{sinc}^2(\pi/q) \qquad (4\text{-}46)$$

in which q = the number of elements per 2π delay period. This may be recognized as the [Fourier transform]^2 of a uniformly illuminated linear aperture, as described at the end of Section 2.4.1, expressed by Equation 2-2, and illustrated in Figure 2.1. Inserting values of $q = 4$ and 8, Equation 4-46 yields $\eta_4 = 0.81$ and $\eta_8 = 0.95$, respectively, as indicated above. Lower efficiency due to reduced q represents depletion of the main lobe to the sidelobes caused by disruptive wavefront staircasing.

Typical liquid crystal phase retarder elements exhibit a unique loss factor established by the minimum space required to relax its orientation from a 2π phase shift to zero. This "flyback" transition is analogous to the flyback time τ of many conventional scanners (including acoustooptic), discussed in Section 4.3.2, Duty Cycle, and expressed in Equation 4-1 as $\eta = 1 - \tau/T$, where T is the full scan period. Note that this represents a time loss, affecting among other things the bandwidth for a required overall transfer time. In the case of liquid crystal elements, the time terms are replaced with small λ representing the flyback width and Λ the full 2π width, yielding the duty cycle

$$\eta_\Lambda = (1 - \lambda/\Lambda)^2 \qquad (4\text{-}47)$$

squared to denote the radiated intensity rather than the time. The topic of efficiency merits attention for at least two other parameters, the fill factor and the overall vignetting factor. The fill factor accounts for the practical burden of creating a refractive or reflective cell having an operating portion that occupies its full allotted area. The vignetting factor considers the typical loss of input illumination beyond the boundary of the array. In context of power throughput, the final efficiency is the product of the individual factors.

The far field angular beamwidth Θ_B is expressed as a minor variation to the diffraction relation of Equation 2-16, which accounts for the distribution of the input illumination upon the full aperture width Nd and is given by

$$\Theta_B = a\lambda/Nd \cos\Theta_o \qquad (4\text{-}48)$$

in which a is the aperture shape factor modifying the beamwidth, as discussed in Sections 2.1.2 and 3.2 regarding spot size and scanned resolution. Because the output beam angle Θ_o seldom exceeds $10°$, its cosine renders a negligible effect on the beamwidth, yielding a more concise expression, as is that of Equation 2-16

$$\Theta_B = a\lambda/D \qquad (4\text{-}48a)$$

in which $D = Nd$ is the is the full aperture width.

4.10.1.2 Resolution of Phased Arrays

Equation 4-48a and its relationship to scanned resolution (Chapter 3) motivate expression of a topic that seldom appears explicitly in the literature. Although the above equation denotes the output beamwidth—and hence the narrowness or breadth of the principal lobe of radiation—it provides no calibration of the number of steered accessible adjacent spots or lobes that may address the full (one- or two dimensional) field of view. In the context of scanned resolution, we seek this *number of distinguishable elements of information that may be positioned* along a linear track (see Section 3.1.1). In analog form, this scanning process is described by the convolution or shifting function discussed in Section 2.1.3 and illustrated in Figure 2.4. The resolution number is denoted by the letter N, whereas N identifies the number of phase shifting cells in a linear array, as represented relating to Equation 4-48. This is also distinct from typical microwave radar usage of the word *resolution*, which is defined in that discipline as the *smallest increment (finesse)* of beam motion.

As developed in Section 3.1.1 on Scanned Resolution, the basic Equation 3-3 yields

$$N = \frac{\Theta}{\Theta_B} \qquad (4\text{-}49)$$

in which Θ is the full deflected field angle. One-half of this angle is represented by Equation 4-43a as the (positive) first diffracted order of a (blazed) grating (i.e., $n = +1$ in equation 4-43). When the elements in the array are addressed in complementary phase sequence, the same deflection magnitude results in the opposite ($n = -1$) direction. Thus, for typically small Θ_o in Equation 4-43b, and with $D = Nd$, we form the numerator for Equation 4-49 as

$$\Theta = \frac{2\lambda}{qd} = \frac{2\lambda N}{qD} \qquad (4\text{-}50)$$

With Equation 4-48a providing the denominator of Equation 4-49 and adding the one boresight position, the steered resolution reduces to*

$$N = \frac{2}{a}\left(\frac{N}{q}\right) + 1 \qquad (4\text{-}51)$$

independent of the wavelength. Setting aside temporarily the constant $2/a$, the ratio of the two variables N and q—that is, *the total number of elements divided by the number of elements per phase reset* dominates. Thus *the number of phase resets in the array* is the principal variable that determines the steered resolution (accessible spots) of a one-dimensional phased array. Although the full width of the array D establishes the narrowness of the radiated lobe, the number of such adjacent lobes within the field is expressed more directly dependent on N/q. (Note: $N/q = D/\Lambda$, where Λ is the array period.) Because q is not constant, this also affects how close the adjacent steering states can be [Wat2].

Equation 4-51 includes the aperture shape factor a, typically a constant of the system. When assumed of value one, it denotes uniform illumination upon a rectangular aperture (in one dimension). This yields the far-field intensity distribution of the $\text{sinc}^2(x)$ function (Equation 2-2b, Fig. 2.1), having a main lobe within equispaced null intervals. Rayleigh resolution requires this uniform illumination on a rectangular aperture and, further, that the adjacent spots in the far field overlap such that the maximum of each main lobe coincides with the first null of each adjacent spot. Further delimiting Equation 4-51, it is impractical to form a modulo 2π array in which q is less than three cells, in view of the resulting disruption of the ramp wavefronts and the loss in efficiency. Thus, letting $a = 1$ and seting $q_{min} = 3$, steered resolution is sometimes expressed as an assumed Rayleigh resolution with $q = 3$, forming

$$N = 2(N/3) + 1 \qquad (4\text{-}51a)$$

---

* Although Equation 4-51 appears to depart from the fundamental Equation 3-5 for scanned resolution $N = \Theta D/a\lambda$, this Θ is established by the diffractive structure of the array and *is not an independent variable*. Accordingly, substituting Equation 4-50 into the classic Equation 3-5 and adding one for the boresight forms this Equation 4-51 directly: $N = [(2/a)(N/q)] + 1$.

Depending on aperture illumination, the value of a may differ from 1, and N and q may vary to accomodate spot efficiency and spot steering. A familiar illumination is the Gaussian function, with adjustment of the overfill to make the intensity more uniform across the full aperture width D. The degree of overfill is, however, moderated by the reduction in light throughput due to the loss of illumination beyond the array boundary (vignetting). It is also balanced by the appearance in the far field of fine structure beyond the main lobe, approaching the appearance of the sinc^2 function when D is illuminated uniformly.

To quantify this value of a, we consider a different condition than that which generated Table 3.1 in Chapter 3. The left side of that table summarizes the value of a for the Gaussian beam of nominally round cross section falling either completely within the deflecting aperture ("untruncated" column) or when the $1/e^2$ intensity of the input Gaussian beam occurs at the aperture boundary ("truncated" column). These are the two prominent conditions of illumination of most conventional deflectors. Some, such as the acoustooptic devices, exhibit a rectangular cross section of width >> height, meriting a different evaluation of the shape factor a. The consequences of illumination with a beam that is Gaussian primarily in the D-direction is evaluated and summarized [Bei2], providing data of current interest.

These data are for the variable beam width W (Gaussian at $1/e^2$ intensity) illuminating the constant width D of a linear phased array. Assigning a parameter $\rho = D/W$, when $\rho = 1$ then $W = D$, whereby the $1/e^2$ input beamwidth matches the aperture width D. At $\rho = 1.5$, the array aperture is 1.5 times wider than the $1/e^2$ beamwidth. In this condition, Figure 2.2 indicates that the aperture delimits the Gaussian function at $\pm3\sigma$, where its intensity has tapered off to a very small fraction of the maximum value. This represents a practical limit on the narrowness of the illuminating beam. At the other extreme, when $\rho \to 0$, this requires the input beamwidth $W >> D$, corresponding to the condition of extracting near-uniform illumination from the center of the Gaussian beam and encountering extreme light loss beyond the aperture. This is the case of $a = 1$. The aperture shape factors for the other one-dimensional gaussian cases are determined [Bei2] and given for $\rho = 1$, $a = 1.15$ and for $\rho = 1.5$, $a = 1.35$.

Related to the topic of resolution is consideration of the smallest increment of beam position (finesse) addressable by a phased array. Returning to the basic wedge deflector of Figure 4.29a, one can reduce its wedge angle or its refractive index to provide any small increment of refracted angle, consistent with practical manufacturing tolerances. Figure 4.29c shows, however, the periodic patterning of a phased array requiring a quantized cellular arrangement. Thus minimal beam shifts are limited by integral changes in the phase resets. Equation 4-43a defines the small diffracted angle in one direction as $\Theta_o = \lambda/qd$. Thus increasing q (delay elements per phase reset) by one cell narrows the deflection angle incrementally. Adding one cell to each group reduces the angle to $(q/q+1)$ of its initial value. The higher the initial q number, the finer the change in angle. Because it is necessary to form the array with full 2π resets, fractional residues can be filled by approximating uniform spacings with a few alternating $(q), (q+1), (q), (q+1)$ periods and programming all to form 2π resets. This minimizes wavefront distortion. Extending this alternating procedure to a larger portion of the array length—not just to equalize a residue—permits achievement of a smaller angular change, especially when the array is composed of groups having a low q number. Not yet evaluated is the possible degradation in efficiency with application of this procedure. Any degradation is likely less significant for groups of high q number.

4.10.1.3 *Phased Array Developments*   After the pioneering work identified in Section 4.10.1, more recent development merits review. Figure 4.29c illustrates an array composed of bulk electrooptic crystals of lithium tantalate, which in the reference by Mey require a significant axial length to appreciate 2π phase retardation with reasonable applied voltage. A resonant approach utilizing a compact stack of alternating GaAs and AlAs dielectric layers was reported in 1999 [Keys] having potential for operation in transmission and in reflection. Thirty-period stacks of approximately 5-mm length operating at a selected wavelength are shown to approach full 2π phase delay with some sacrifice in output amplitude stability. Design improvements are discussed.

Work using nematic-phase liquid crystal electrooptic retarders is detailed comprehensively in a 1993 Air Force document [Dor]. An effective description is available in more accessible form [McM]. The types of material are known as E7 and PTTP-33 liquid crystals, having birefringence $\Delta = (n_e - n_o) \simeq 0.2$ in the infrared. Thus a cell need be only 5 optical waves thick for a full-wave phase shift in transmission and only 2.5 waves thick in reflection. The thinner the cell, the shorter the time for molecular reorientation. Switching speeds in the millisecond range with high efficiency diffraction-limited steering have been demonstrated at 10.6 μm with CO_2 lasers and at 1.06 μm and 0.53 μm with Nd:YAG lasers. Further development tested tandem scanners as a means for adding the contributions of two or more deflectors, each in its optimal operating range. Thin one-dimensional arrays having crossed electrode patterns may be summed, one for azimuth and one for elevation. Or individual deflectors requiring excessive spatial separation may be cascaded [Bei2] using relay optics represented by Figure 1.7 to avoid walk-off of the beam from the second aperture by the action of the first deflector. This is analogous to the familiar woofer-tweeter expansion of acoustic bandwidth. In a major tested system, a course (large angle) deflector was summed with a fine (small angle) one, multiplying the number of steering states of each. A group of discrete large angular positions of the course deflector were filled with additional numbers of small steered positions of the fine unit. Electrical connection was implemented in an elegant leadout arrangement termed "multiple-state" architecture, which reduced substantially the number of wires to be addressed to 768, whereas 40,000 would have been required for a single fully addressable array having the same number of addressable beam directions.

Another approach to tandem arrays [Tho], called the discrete/offset bias cascade, reduces potential "noise" (beam artifacts) in the instances of large quantization mismatches when cascading phase-delayed groups. It also reduces significantly the number of control lines that would be required to provide the same resolution from a single phased array. This was accomplished by programming a typical first array of cells of width d arranged in a regular Λ-group fashion (called "discrete scanning") and directing the resulting wave pattern into a second stage (called "offset phases") composed of wider cells of width $\Lambda = qd$ such that they register over the Λ-groups of the first stage. This second stage adds constant phase delays to the wavefront segments from the first array, to form smoothly joined wavefronts. Experiment demonstrated improved overall diffraction efficiency, along with the use of a reduced number of control lines. A similar approach was demonstrated with microlens arrays [Flo] (Section 4.10.2). Also significant in this work is the use of an electrooptic phase retarder other than liquid crystal. The material selected is PLZT (lead lanthanum zirconate titanate), exhibiting a large electrooptic coefficient, broadband optical transmission, very fast switching, and good thermal stability [Hae]. This well-documented ceramic material is familiar in electrooptic modulator and deflector applications.

Mirrored pistonlike phase adjustment has been reviewed [W&M], and later work [Burn] describes both continuous optical phase change and fixed (binary) phase shift. When continuous, the height of the array mirror is decreased by analog electrostatic attraction toward the substrate. In binary operation, a larger fixed voltage pulls the element to the substrate over a fixed distance. The fabrication constraints are stringent, considering its dimensional characteristics, even with application of advanced surface micromachining technology and materials. The first generation device was a linear array of 128 mirrored elements; each 27 μm wide and 110 μm long and held 2 μm above the substrate by end flexures. Residual stress curved the (intended flat) mirrors to sag depths of approximately 10 nm and 44 nm along the width and length, respectively. Control voltage to ≈20 V provides continuous phase shifts to a full 2π change. Beyond 24.5 V, the mirrored element executes "snap-through" to the substrate, forming the binary condition. Optical efficiency was low, because of poor fill factor of the 19-μm-wide gold reflector in a 30-μm array period and by destructive reflected interference between the gold coating and exposed polysilicon borders. Planned improvements estimate raising maximum efficiencies to 63%.

Problems in broadband operation of phased arrays have been reviewed [McM], and work has been directed toward their solution [Wat3, Sto]. A wavelength-independent phase shift is achieved by polarization modulation of chiral smectic liquid crystals (CSLC), providing action similar to the mechanical rotation of a waveplate. However, grating dispersion remains because of wavelength deviation from nominal 2π phase resets, rendering a variation in efficiency η_d similar to Equation 4-46

$$\eta_d = \text{sinc}^2(\varepsilon) \qquad (4\text{-}52)$$

where ε is the chromatic error due to mismatch of the nominal 2π phase reset. When in error, not only is energy lost, but sidelobe amplitudes increase and nondiffracted components result in image blurring and interference from sources outside the desired acceptance angle. This dispersion is nulled with application of special achromatic optics [M&Z] to restore the system to single-focus imaging. Additional work using chiral smectic-A ferroelectric liquid crystals (CSFLC) is expressed [Dro] to provide phased arrays with fast switching (1–100 μs) and a low drive voltage requirement (5–10 V).

4.10.2 Decentered Microlens Arrays

An alternate to phased array beam steering is the decentering of a group of lenses with respect to a matching lens group. Although the fundamental beam steering action of an individual pair of decentered lenses differs uniquely from the phased cellular systems described above, when lenses are assembled into mating periodic arrays, the combination exhibits some of the basic characteristics of phased arrays, including the functioning as blazed gratings [Wat1].

Consider Figure 4.30*a* illustrating a pair of lenses (1 and 2) oriented originally afocal and then decentered from their common axis through distance Δ (dotted axes). Whereas the input beam at the left focuses lens 1 to an initially common focal point, its diverging beam continues into lens 2 shifted off its axis, resulting in deflecting the recollimated output beam through the angle Θ_o. This may be confirmed with a simple ray trace. Thus a transverse shift of lens 2 with respect to lens 1 affects beam steering. The vignetting of the output beam and the related diversion of the residual output flux outside the lens are discussed below. Major constraints to using this basically simple two-lens technique are its limitation on the width of the lens aperture, consistent with the energy requirement for sufficient Δ-shift within reasonable burdens of acceleration of massive components. These limitations represent, in fact, some of the reasons for seeking agile beam steering.

However, consider miniaturizing many lenses 1 and 2 (maintaining the f-number), formatting them into arrays of microlenses, per Figure 4.30*b*, and illuminating the group from the left by a single broad beam. The steered waves sum into the total field in a manner similar to those of the prior phased arrays. Immediate consequences are a significant decrease in mass for a given overall aperture size (similar to Fig. 4.29*b*) and a decrease in shift distance Δ required to steer the beam. The effect is a dramatic reduction in array travel and in the acceleration/deceleration forces required for rapid beam steering. Although the steered wavefront is discontinuous, the periodic output components exhibit the characteristics of a blazed grating. The array of steered beams (rays) corresponds to normal wavefront segments that are tipped at the same slopes. When, at the operating wavelength, their junctions exhibit 2π phase differentials, they form the sawtooth pattern typified by a blazed grating, providing high diffraction efficiency.

The technique of Fig. 4.30*b* is satisfactory for small steered angles, where the "spurious" components remain a small residue of the desired output (maintaining high fill factors at the second lens array). However, at wider steering angles, when the vignetting and the disruptive effects of the spurious components become significant, remedies are considered. A classic method for the control of vignetting is the use of a field lens [Lev], introduced into the microlens array [Wat1] and adapted to proper lens movement. Figure 4.30*c* illustrates this as a variation to Figure 4.30*a* with a field lens (FL) inserted at the focal crossover of the original lens positions 1 and 2. The bar over the output pair of lenses represents physical connection for simultaneous motion. With equal focal lengths of all lenses, the expanding light cone fills completely lens 2 through Δ-shift of the lens pair, readily confirmed with a ray trace.

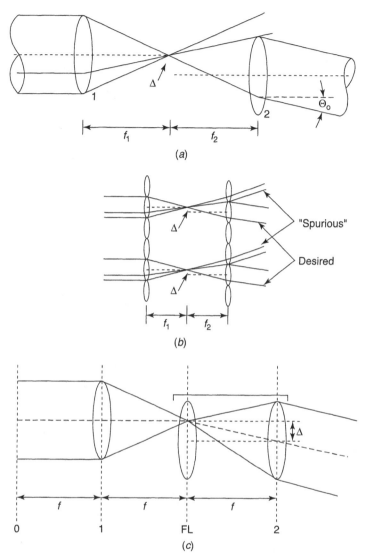

Fig. 4.30 Beam steering with decentered lenses. Original afocal lenses displaced through distance Δ. [*a*] Macroscopic lens pair, showing Δ-shift deflecting output beam through angle Θ_o, while upper portion of the beam bypasses lens 2. [*b*] Arrays of microlenses performing as in [*a*], but lighter and with smaller Δ-shift. The desired components accumulate while the bypass portions are directed into a spurious angle. [*c*] Field lens form of [*a*], where (FL) provides complete filling of lens 2. When in [*b*], this synthesizes the output wavefront of a blazed grating. After E.A. Watson, "Analysis of beam steering with decentered microlens arrays," in *Optical Engineering*, Vol. 32 (1993). Reproduced by permission of the publisher.

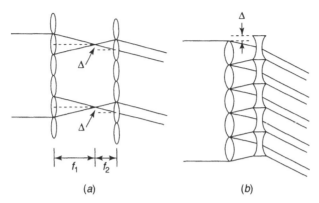

(a)   (b)

Fig. 4.31 Increase of focal length ratio f_1/f_2 to eliminate spurious components over a range of operation. [a] Similar to Fig. 4-30b (Keplerian form) with $f_1/f_2 \simeq 2$. [b] Analogous Galilean form with $f_1/f_2 \simeq 2.5$. Although the compressed beam energy is conserved, the output wavefront represents that from a discontiunous blazed grating. After E.A. Watson, "Analysis of beam steering with decentered microlens arrays," in *Optical Engineering*, Vol. 32 (1993) and W. Goltsos and M. Holz, "Agile beam steering using binary optics microlens arrays," in *Optical Engineering*, Vol. 29 (1990). Reproduced by permission of the publisher.

This technique for sustenance of output efficiency and spectral quality is directly transferrable to the microlens phased array of Figure 4.30b with an added plane of field lenses affixed to the output array. The extra inertia can be accommodated by the force of piezoelectric or electrodynamic drive transducers. Or the single element may be moved, instead. A microlens-field lens design was fabricated and measured [M&W] over a ±1.6° field, for use in optical data storage. Larger angles (±17°) have been demonstrated [Wat4], but with loss of beam quality.

Alternate considerations for suppression of the spurious beams during Δ-shift are represented in Figure 4.31. The method of Figure 4.31a is introduced [Wat1] to illustrate partial tolerance for beam displacement on lens 2 by changing the ratio of focal lengths. The initial condition of $f_1 = f_2$ is adjusted to $f_1/f_2 > 1$. This forms a modified beam compressor (see Fig. 1.7) with a compression ratio $\simeq 2:1$. A similar approach is indicated [G&H] using a positive-negative lens combination. The Figure 4.31a method employs the equivalent of an afocal Keplerian telescope and the Figure 4.31b method that of a Galilean telescope. Although the spurious components of Figure 4.30b may be abated over its initial range of operation, the fill factor at the second array is reduced significantly. Although the energy is conserved in this reduced but integral light cone, the ideal sawtooth pattern of the blazed grating is disrupted dramatically by the truncated sawtooth function.

This, in turn, causes its own spurious noise [Wat1, G&H] which limits operation to a small range of Δ-shift. It is proposed, therefore [G&H], that the second array be maximally filled, ideally by reducing the lens separation in Figure 4.31*b* to zero. This is approached with the development of thin binary optics microlens arrays.

Binary optics arrays are fabricated by utilizing high-resolution etching and transfer techniques having high finesse to form binary representations and variations of Fresnel zone patterns on substrate materials. One hundred percent fill factors of lenslet arrays are attainable with matching and abutting lens shapes (e.g., hexagonal). Imparting a multilevel relief structure approximates a continuous phase profile in a stepwise manner, allowing achievement of high diffraction efficiency. As presented above for a phased array composed of q elements per 2π phase reset (Eq. 4-46), similarly, the efficiency η_b of a multilevel binary optic [Swa] of m levels within one width of a Fresnel feature is given by

$$\eta_b = \mathrm{sinc}^2(1/m) \qquad (4\text{-}53)$$

Three etching steps forms eight levels, yielding the theoretical efficiency of 95%. An experimental system [G&H] utilized such arrays of f/5 microlenses, each of 0.2-mm diameter (having focal lengths just short of 1 mm) in a hexagonal grid array. The second (negative) lenslet array was spaced from the first by 10 μm, allowing relative translation in two dimensions. Three masks were used to form the eight-level approximation of a parabolic phase profile (which allows steering without angle-dependent aberrations). This system steered a 6-mm HeNe test beam over an 11.5° field using ±0.1-mm travel at a 35-Hz sweep rate. Practical mask alignment, etch, and transfer errors during fabrication reduced the 95% maximum efficiency to measurements of 84% and 72% for the positive and negative lens arrays, respectively. Overall throughput efficiency of the unsteered beam measured approximately 50%. The f/5 system exhibits low efficiency when steered. This is expected to improve with fabrication and operation at lower f-numbers. Although the accuracy of the utilized engraving system was evaluated at 0.1 μm, cumulative fabrication errors in the present work were estimated to limit attainable minimum feature size to ≈0.4 μm.

A variation on the above work was conducted using "phased-array-like" binary optics [Farn]. The change is effective principally during Δ-shift, when a complementary pair of lenslike elements (as in [G&H]) renders incomplete filling of the lens aperture with the desired component. This results in each lenslet pair forming only a portion of the ideal phase ramp for a proper blaze, leaving a residue from each lenslet pair ramped in the wrong direction. In the basic phased-array-like method, the continuous quadratic phase function (parabolic per [G&H]) is sampled at equal intervals of Δx, forming a stepwise (piecewise flat) matching of the continuous phase profile. The Δ-shifts are then conducted in integral increments of Δx, to $n\Delta x$ for diffraction to the nth order. Because the flat segments in the residue region maintain modulo 2π phase differentials with respect to an idealized linear continuation, this formerly disruptive region is shown analytically to render a continuous linear phase profile across the full aperture.

Experimental binary microoptics were designed to compare measurements of phased-array-like and microlens arrays. They were fabricated simultaneously and adjacent on the same thin quartz substrate. Tests confirmed that the phased-array-like structure provided a significant ($\approx$50%) increase in intensity in the steered mode and exhibited less than 1% leakage into its immediate (local) sidelobes. Although stronger distant sidelobes developed from the phased array, they were well separated from the steered mode, by $\pm$64 mode orders in these tests.

4.10.3   Summary of Agile Beam Steering

This section has concentrated on two dominant approaches to very low-inertia laser beam steering: the phased array and the decentered lens array. Although their operating principles appear to differ, it is shown that they both form diffraction gratings yielding output wavefronts having the characteristic of a blazed grating. The result is high diffraction efficiency in a selected diffraction order in one direction of a symmetric pair of ($\pm$) sets. The complementary set is provided by programming the negative action of the first, similar, for example, to mirror deflection within a zero-balanced restoring force, requiring (+) and (−) excitation to steer through the full available field.

Some distinctions between and characteristics of the two array techniques are noteworthy. The dominant phased array method is completely electrooptic, whereas the microlens array requires very small displacements, with concomitant low inertia acceleration/deceleration. An alternate phased array utilizes minute ($\leq\lambda/4$) axial displacement of individual micromirrors. However, it encounters difficulty in fabrication to high optical integrity and in achieving an optical fill factor that approaches that of the electrooptic type. Alternate microlens arrays are formed of Fresnel lens-type binary optics. For low f-number lenticules, whose theoretical steered efficiency is high, their minimum feature sizes become miniscule and also, thus far, difficult to fabricate.

Phased arrays utilize complex multielement electrical programming, whereas the lens array systems require controlled and very precise positioning of the lens assembly over variable small distances. These associated operations can impose significant burdens of mass, volume, and cost of the auxiliary facilities required for address and position control. Although this factor is generally not detailed in the literature, a comparative analysis [McD] indicated several related observations regarding the 1995 state of the art. The authors preferred a microlens array over the liquid crystal phased array, avoiding thereby the "heavy burden" on electronic control of the many individual phase-delay elements. An aplanatic field lens-type (3-lens) array was designed, having a 6-mm path length through a 1-mm aperture of silicon lenslets. Along with x and y translation, z-axis motion was programmed to minimize aberrations. Image quality was within 1.3x diffraction limit. The principal comparison was of this system with respect to a two-galvanometer set, assembled of high-quality commercially available components. Detailed evaluations confirmed that the microlens system steered faster, consumed lower power, and packaged much smaller and ligher. However, no comment appeared on minimizing the scanning mirror sizes and inertia and the mass of the mechanical assembly. Nor was the use of relay optics indicated, to allow the second galvanometer to be as small as the first, to render a dramatic reduction in mass and inertia. Nor was the adaptation of the single articulated mirror that is driven angularly in two dimensions [New, Ball, Ber] expressed. Although some alternatives may exhibit negative trade-offs, these evaluations (including auxiliary control facilities) are very useful in the context of the relative capabilities for meeting system requirements.

CHAPTER 5

CONTROL OF SCANNER BEAM MISPLACEMENT

Two forms of beam misplacments are discussed here. The first, known as "cross-scan" error, results in small shifts of the beam in that direction that can cause a perceptible defect called "banding." This typically develops from small angular errors of the rotating shaft (wobble) and/or from angular nonuniformities of polygon facets in the cross-scan direction that displace the beams during successive scans. Extensive research has been conducted to relieve this problem without the burden of stringent mechanical tolerances, resulting in effective control methods. Attention is devoted here to these successful options.

The second form is a less frequently encountered beam misplacement that forms a "ghost image." In contrast to the above-described source of error, this effect does not result from a mechanical deviation of polygon facets or from a flawed lens design. Rather, it is a consequence of the propagation paths of the input and scanned beams and of the redirection and refocusing of the scatter from the original focused beam that is incident properly on the storage medium. Two effective correction methods are discussed.

An unexpected interrelationship between these two entirely different types of beam misplacements is introduced, which appears between one of the procedures for ghost abatement and one for the above-noted cross-scan error. An excellent alternative eliminates the ghost image while avoiding conflict with the control of cross-scan error.

5.1 CROSS-SCAN ERROR AND ITS CORRECTION

The discussion of beam deflection in Chapter 4 identified some imper-fectons in the positional accuracy of several scanning devices, along with an indication of some control methods (Sections 4.2, 4.3, 4.3.1, 4.4.1.1, 4.5.3, 4.5.4, and 4.6.3 and Table 4.1). This reiteration serves not only to introduce the topic of this chapter, but to highlight its opera-tional significance. The requirements for optical scanning transcend the basic need for efficient beam positioning by seeking its accomplishment with high positional integrity. Although almost all scanning devices execute beam deflection along one axis (the "along-scan" direction), and may be accompanied by positional errors that require correction in that direction, it is the quadrature axis ("cross-scan" direction) that is most difficult to access, hence most insidious and awkward to control. After exercising reasonable care in minimizing scan perturbations, the along-scan direction benefits from the basic property of being time related, allowing complementation of the residual scan nonuniformi-ties with timely indexing of the data stream. A classic method of accu-rate tracking is to derive a synchronous auxiliary or pilot beam from the same deflector, scan it across a precise grid, detect the equally timed impulses, and then trigger the signal data stream accordingly to provide a corresponding spatial integrity [Toy]. Various related techniques are instituted to take advantage of this time-space relationship. A preva-lent example is the precise initiation of data scans that are triggered by "start-of-scan" (SOS) pulses that are detected optically from the actual or auxiliary deflected beam.

Because the cross-scan errors appear in quadrature to the along-scan errors, entirely different correction methods must be considered and instituted. Depending on their sources and periodicity (which may be pseudorandom), the defects may be short or long fractions of the scan period. When complete scan lines are misplaced, this formation appears often as the very perceptible and therefore insidious problem called "banding." The intensive effort that is directed to alleviate cross-scan errors has achieved substantive success. This section concentrates on these error-reduction disciplines and techniques.

5.1.1 General Considerations and Available Methods

The high-performance rotational scanner (having deflecting mirrors or holograms) depends on precise shaft orientation about its axis. The angular uniformity of these elements with respect to the shaft axis and of the axis with respect to its frame imposes substantial demand on fab-

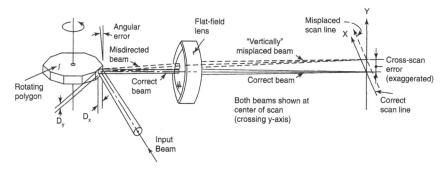

Fig. 5.1 Rotating polygon scanner having angular error (exaggerated) that misplaces the scanned line in the y-direction. Similar error results from shaft, bearing, or balance perturbations. From L. Beiser, *Laser Scanning Notebook*, SPIE Press (1992). Reproduced by permission of the publisher.

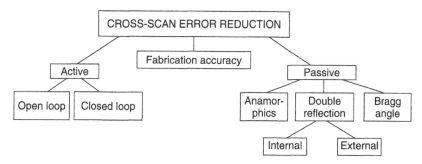

Fig. 5.2 Resources and techniques for cross-scan error reduction. From L. Beiser, *Laser Scanning Notebook*, SPIE Press (1992). Reproduced by permission of the publisher.

rication procedures and on the consequential cost. Figure 5.1 illustrates the (exaggerated) beam misplacement of a typical polygon scanner due to its angular error in the cross-scan direction.

A composite of resources for cross-scan error reduction is represented in Figure 5.2. *Fabrication accuracy* may serve as the only (and costly) discipline, or it may be augmented by any of the alternate methods. The *active* methods utilize high-speed low-inertia acoustooptic, electrooptic, or piezoelectric deflectors [Bei2, Don] or lower-speed galvanometers that are programmed to steer the beam to rectify its positional errors. Although open-loop methods may be adequate for repetitive errors, elegant closed-loop methods are often required to correct pseudorandom perturbations. The complexity and cost of this active approach must be compared to the alternatives of increased fabrication accuracy and of the use of *passive* techniques.

5.1.2 Passive Methods

Operating continuously and automatically, passive techniques require no programming. Utilizing optical principles in novel configurations, they reduce beam misplacement due to angular error in reflection or diffraction. Error reduction of tilted holographic deflectors operating near the Bragg angle is represented by Equation 4-15 and is discussed in Section 4.4.1.1. Two additional passive techniques are now described that employ anamorphic optics and the principles of double reflection.

5.1.2.1 Anamorphic Error Control The most prominent treatment, anamorphics, may be applied to almost any deflector. The basics and operational characteristics [Bei1] are summarized here. Anamorphic optics exhibit unequal power along quadrature meridians. A cylindrical lens, for example, exhibits power in one direction only.

Separating the angular resolution (N of Equation 3-5) into quadrature (x, y) components and denoting the critical cross-scan error direction as y, the resolvable error is expressed as the number of misplaced resolution elements

$$N_y = \frac{\Theta_y D_y}{a\lambda} \qquad (5\text{-}1)$$

typically, as a fraction of the height of a single element for such small errors, in which $a\lambda$ is assumed constant, Θ_y is the *causative angular error* of the misdirected beam in the cross-scan direction (Fig. 5.1) and D_y is the *height* of the beam illuminating the deflector. The objective is to make $N_y \rightarrow 0$. [$N_x = \text{f}(D_x)$ represent the *desired* along-scan resolution, which is nominally unaffected by these operations.]

The inaccuracies of fabrication and rotational dynamics determine Θ_y (the beam error angle). To reduce N_y, anamorphics are introduced to reduce D_y proportionately, per Equation 5-1, as required for the magnitude of error reduction. This is usually accomplished as illustrated in Figure 5.3, with a first cylindrical lens focusing the illuminating beam in the y-direction upon the deflector. (D_x remains unaffected.) As D_y is reduced to D_y', so is the y-component error, per Equation 5-1. After deflection, the y-direction beam height is restored with complementary anamorphics (a second cylindrical element), reestablishing via the flat-field lens the nominal converging beam angle to form the nominal focused spot height. The x-direction characteristics remain (ideally) unaffected. Following Equation 5-1, the y-error is reduced to the ratio

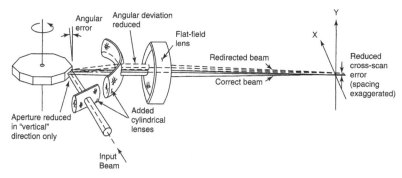

Fig. 5.3 Input beam compressed in height with a first cylindrical lens that reduces D_y and cross-scan error proportionately. A second (toroidal) cylindrical lens restores the beam to form an (ideally) unmodified spot. The second cylinder may be integral within the flat-field lens. From L. Beiser, *Laser Scanning Notebook*, SPIE Press (1992). Reproduced by permission of the publisher.

$$R = D'_y / D_y. \qquad (5\text{-}2)$$

where D'_y is the compressed beam height on the deflector and D_y is the original beam height on the deflector* [Bei3].

* The above analytic approach and its consequence that yields Equation 5-2 departs fundamentally from the more familiar published descriptions, which adapt the concept of the original error correction technique [Fle]. In that principle, the first cylinder in Figure 5.3 converges the input beam to effectively zero height on the deflector facet. (The resulting very thin line remains fixed with respect to the optical axis during any cross-scan error shift of the facet.) The near-zero height on the deflector is then, ideally, reimaged conjugately by the flat-field lens onto the image plane with zero perturbation. This is the limit condition of the above analysis, requiring the error $N_y = 0$ in Equation 5-1, whence the remaining variable $D'_y = 0$.

Although a valid concept, reduction of the beam height on the facet to zero does not allow for adjusting the magnitude of correction to match the operational requirement for error control. In practice, aberrating effects can develop because of excessive reduction of D_y by the converging "wedge" beam. During polygon rotation, the facet shifts its extent into and its angular relationship to the optical axis. This can change the position and degree of focus of the wedge beam on the facet and effectively modify the defocus uniformity with alternating polarity during shaft rotation. If the effect is sufficient, the conjugate image is thereby enlarged and aberrated dynamically. It is judicious, therefore, to institute control *only to the degree required* to correct the residual *perceptible* cross-scan error.

Equation 5-2 of the above analysis provides for such adjustable correction. It may be implemented by selecting the focal length (and location) of the first anamorph to provide the f-number (F) for the required $D'_y \approx F\lambda$ (per Equation 2-25) and then instituting a complementary anamorphic adjustment in the output optics. An effective alternative is to beam-compress and beam-expand anamorphically telescopically (per Fig. 1.7 paraxially or with prisms), so that the compressed beam remains collimated in the y-direction. This will negate the indicated aberrations in the image that results from distortion of the intended line-object on the facet during rotation of the polygon.

A variety of anamorphic configurations have been instituted, with principal variations of the second anamorph integrated within the (usually "flat field") objective lens, to reestablish the nominal converging beam angle and to maintain focused spot quality and uniformity during scan.

5.1.2.2 Double-Reflection Error Control In double reflection, the deflector that creates a cross-scan error is reilluminated by the beam in complementary phase to null the error. This is discussed first as applied to monogon-type configurations, followed by its application in polygons. Double reflection can be conducted in two forms, internal and external.

An internal double-reflection scanner is exemplified by the pentaprism of Figure 5.4*a* [Sta]. The shaft is mechanically coupled to the prism. This is an optically stabilized alternative to the 45° monogon of Figures 1.9 and 4.7, operating preobjective in collimated light. Tippling the pentaprism cross-scan leaves the orientation of the output beam unaffected. (A resulting minute *translation* of the collimated output beam is nulled when the beam is focused by the subsequent objective lens.) The pentamirror of Figure 5.4*b* having the same ray path requires significant balancing and support of the unbalanced mirrors, for a shift (e.g., inertial) in the nominal 45° included angle doubles this error in the output beam.

A stable and simpler double reflector is the open mirror monogon [Bei9] of Figure 5.4*c*. Its nominal 135° angle (which is structurally rigid) serves identically to maintain a constant (90°) output beam angle, independently of cross-scan wobble of the deflector.

Two variations that provide two scans per cycle, as would a two-faced pyramidal polygon or ax-blade scanner [B&J] appear in Figure 5.5. Figure 5.5*a* is effectively two pentamirrors forming a "butterfly" scanner [Mar2], and Figure 5.5*b* is effectively a pair of open mirrors [Bei8]. The included angles of each half-section must be made equal to $\frac{1}{2}$ of the allowed error in the output beam. Also, the center section of method (a) must be angularly stable (e.g., during rotation) to within $\frac{1}{4}$ of the allowed error, for an increase in included angle on one side forms a corresponding decrease on the other. Dynamic considerations include transverse beam displacements as the device rotates, increasing required mirror widths in proportion to the radial distance of the input beam from the rotating axis. This increases further the bulk of method (a) and the need to stabilize the central section against minute angular shifts.

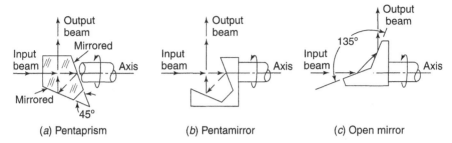

Fig. 5.4 Monogon-type scanners (typified by Figs. 1.9 and 1.10) employing double reflection to null cross-scan error. From L. Beiser, *Laser Scanning Notebook*, SPIE Press (1992). Reproduced by permission of the publisher.

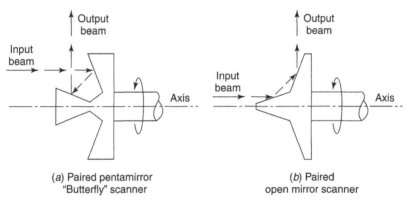

Fig. 5.5 Paired scanners to double the scan rate, employing double reflection to null the cross-scan error. From L. Beiser, *Laser Scanning Notebook*, SPIE Press (1992). Reproduced by permission of the publisher.

The need for near-equality of the included angles of the techniques of Figure 5.5 can be eliminated by transferring the accuracy requirement to a fixed *external* element that redirects the recurrent beam scans [H&S]. One configuration of "external" cross-scan error correction is illustrated in Figure 5.6, shown in the undeflected position. A prismatic polygon illuminated with a collimated beam of required width *D* deflects the beam (only principal rays shown); first to a Porro prism [Lev] or roof mirror, which returns the beam to the same facet for a second deflection toward the (flat field) objective lens. The roof mirror phases the returned beam such as to null the cross-scan error during the second reflection. Several consequences are noteworthy:

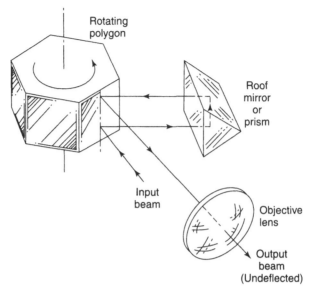

Fig. 5.6 External double reflection to null the cross-scan error. Principal rays shown undeflected, and components and their distances are not to scale. Note consequences of these factors in the text. From L. Beiser, *Laser Scanning Notebook*, SPIE Press (1992). Reproduced by permission of the publisher.

1. The along-scan angle is doubled with the second reflection. Thus the scan magnification $m = 4$ rather than 2.
2. This requires doubling the facets to provide the same scan angle with the same duty cycle. (e.g., Equation 4-5 must be adjusted accordingly.)
3. (a). During polygon rotation, the point of second reflection shifts significantly along the facet and sacrifices duty cycle (e.g., again affecting Equation 4.5).

   (b). Because error nulling is based on the second reflection exhibiting the identical perturbation as the first, the mirror surfaces at the both reflections must convey essentially identical (cross scan) errors.
4. The pupil distance from the (flat field) objective lens ("P" in Fig. 4.2) is extended significantly by the extra reflections and spacings, requiring a larger (and more costly) lens to avoid vignetting.
5. The roof mirror and the objective lens must be positioned to avoid obstruction of the input and scanned beams while minimizing the pupil distance.

With these factors in mind, a detailed layout is essential, combining effectively Figures 4.2 and 5.6 to conclude its viability in the requisite application.

5.2   THE GHOST IMAGE AND ITS ELIMINATION

Under some conditions of writing with a beam scanned by a prismatic polygon, there appears a faint spot on each scan line at a *fixed* location within the useful image field. This "ghost" develops when the scanned focused beam is incident on the storage medium and scatters radiation that is returned, in part, back to an adjacent facet of the polygon. Upon reflection from this facet, it is conveyed forward (through the same optics) and is reimaged as this fixed spot on the storage medium. The continuous exposure of such spots on subsequent scan lines forms a line in the cross-scan direction that can perturb the image. The degree of annoyance depends on the scatter characteristic of the beam incident on the medium, the transmission of the returned and redirected beams, and the gamma (image density vs. exposure) of the storage medium.

A common configuration capable of ghost imaging is illustrated in Figure 3.7, in which the input and scanned beams appear in a plane perpendicular to the rotating axis and all polygon facets are parallel to this axis. At some orientations of the input beam, scatter of the focused spot at the image surface can return through the lens, illuminate the adjacent facet (typically, the one unseen in Fig. 3.7), and be reflected and redirected through the lens, to be refocused at a point that remains at a fixed position on the scan line during polygon rotation.

This fixed beam position, *independent of polygon rotation*, is again a result of the property of *double reflection*, discussed above in this chapter regarding the devices illustrated in Figures 5.4 and 5.6. The effect is identical to that of Figure 5.4 in which the pair of mirrored facets is part of the same substrate that rotates incrementally (in the plane of the paper) as "wobble." Figure 5.4c is most analogous. Let the substrate execute an increment of rotation clockwise. The first reflection is rotated *two* increments clockwise, one for the substrate and one to maintain equal incident and reflected angles, the familiar $m = 2$ factor of scan magnification. This *extra increment* of beam rotation is transferred to the second mirror, *narrowing* the angle to its facet normal and narrowing the reflected angle by the same amount, yielding the same *total* angle, *independent of rotation*. Because the imaging lens in the current case keeps the principal outgoing and incoming rays parallel (discussed below), the rotating polygon, with mirrors *outside*, maintains the ghost output angle constant.

5.2.1   Skew Beam Method of Ghost Elimination

The basic configuration of Figure 3.7 is adaptable to simple correction, with some trade-off. If the input beam is directed to the polygon at a slight skew angle to the plane perpendicular to the rotating axis [Mas] (see Figs. 1.4 and 4.3), the scanned line will be displaced slightly vertically (and also be bowed slightly). Importantly, the doubly reflected ghost spot will then be doubly displaced from the scanned line. When shifted sufficiently, the ghost beam can be blocked just before it reaches the storage medium—a form of schlieren technique. If required, the minor resulting bow (see Section 4.6.2) can be controlled independently.

This method of ghost elimination by cross-scan beam skew is, however, not generally applicable to the laser scanners that use anamorphic correction for cross-scan deflection error (Section 5.1.2.1 and Fig. 5.3). It conflicts with the intended process because anamorphic correction reduces the beam displacement error toward zero at the image plane, frustrating the insertion of a beam block. When only minor correction is required, a sufficiently small anamorphic adjustment may be applied to allow execution of the above skew beam technique. This adjustable cross-scan control capability is discussed in the anamorphic error section above. If the cross-scan error is of sufficient magnitude that it requires a strong correction, or that the resulting bow is inappropriate for correction, a simple and effective different technique for ghost elimination is available.

5.2.2   Beam Offset Method of Ghost Elimination

In contrast to the above method, this procedure maintains all beams in a plane that is perpendicular to the rotating axis of the polygon (as for the mirror of Fig. 1.8). This allows application of anamorphic correction of cross-scan error without the above conflict with ghost beam abatement.

Figure 5.7 illustrates the method. It is discussed first to identify its parameters and then to describe the correction process. A 12-faceted polygon scanner is shown with the facets clarified in the region of interest. This number of facets was selected as typical; the method is general. Other principal components are the imaging lens housing and the image surface. The lens can be conventional, or it may provide flat-fielding ($y = f \cdot \Theta$). Ray paths beyond the lens represent those of a conventional lens ($y = f \cdot \tan\Theta$) to simplify the ray trace. Inside the lens housing is shown the principal plane of the lens elements, whose properties are utilized to render the ray propagation. All rays are the principal rays of beams of finite width (see Fig. 4.2), again, to simplify the illustration.

The input beam is incident on a facet that is so oriented that the reflection propagates to the right along the lens axis. The facet normal is indicated n_o, and the input and reflected rays are disposed $\pm 22.5°$ about this facet normal. Thus the input beam is angled $45°$ to the lens axis. The angle between facets of the 12-facet polygon is $2\pi/12 = 30°$. Hence, the maximum available scan angle is $\pm 30°$. Assuming a duty cycle of $\eta = 50\%$, the working scan angle is $\pm 15°$. The facet above the operating one is identified with normal n_1. It exhibits an opportunity to generate a ghost image that appears *within* the scanned image format. A comprehensive analysis [Mar4] includes ghost image appearance *outside* the scanned format, as generated by this and other facets.

With adjacent facet angles displaced by $2\pi/n$ ($n \equiv$ number of facets), ray 2 is always displaced from the input beam by $4\pi/n$, $60°$ in Figure 5.7. (A "test" broad collimated beam incident to the left on both facets along rays 0 and 1 splits and reflects toward the input beam and along ray 2, separated by $60°$.) By setting the angle of the input beam to the axis to $45°$ ($60° - 15°$), ray 2 directs its beam to the format limit. If narrowed to $<45°$, the ghost beam is positioned *outside* the format, for this $60°$ angle ($4\pi/12$ facets) is constant.

The $15°$ angle is $\frac{1}{2}$ of the $30°$ of available scan angle (one side), limited by the 50% duty cycle. Identifying this angle as θ, then

$$\theta = (2\pi/n)\eta \tag{5-3}$$

Defining $\alpha \equiv$ angle between the input beam and the axis, to place the ghost spot *outside* the image field, $\alpha < 4\pi/n - \theta$. Substituting for θ, $\alpha < 4\pi/n - (2\pi/n)\eta$, which reduces to

$$\alpha < (2\pi/n)(2 - \eta). \tag{5-4}$$

For this 12-faceted polygon at $\eta = 50\%$, $\alpha < (360°/12) \cdot (2 - 0.5)$. Thus $\alpha < 45°$ places the ghost spot out of the image field. Equation 5-4 is determined by action at the polygon and is indpendent of the scan format and lens focal length, as expressed for operation with a flat-field

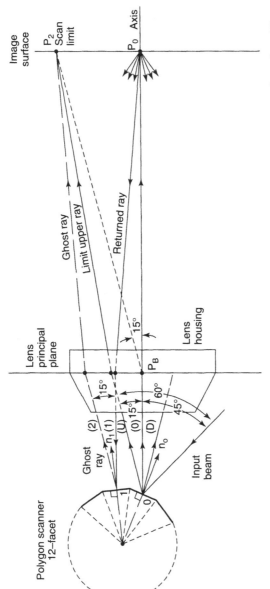

Fig. 5.7 Ghost elimination by beam offset. Only principal rays shown. Input beam is incident on facet 0 at 45° to axis. With polygon in normal position, reflected ray 0, undeflected, continues through lens on axis to image surface at point P_0. Ray of scatter returns through lens to adjacent facet 1 as ray 1, parallel to Ray 0, and is reflected as ghost ray 2 through lens, incident on image surface as point P_2. CCW rotation of polygon deflects input beam along Ray U through lens via limit upper ray, to be incident also at point P_2. An input beam angle of <45° places the ghost image beyond point P_2 securely outside the operating format.

lens [Min]. It is independent of the lens, with minor considerations of flat-field or conventional types. Figure 5.7 illustrates a conventinal lens for the use of standard ray trace procedure. The flat field (or f-Θ) lens places image point P_2 closer to the axis (with intervening points designed to be displaced from the axis in proportion to the scan angle.) Thus the f-Θ lens directs the 15° ray from the polygon to a format limit at slightly less than 15°. The ray trace of Figure 5.7 is detailed below.*

---

* The input beam that is incident on the working facet is reflected as ray 0 and propagates along the lens axis to the image surface at focal point P_0. There, it scatters radiation, a portion of which is returned through the upper part of the lens as ray 1 to the adjacent polygon facet having normal n_1. The forward and return rays (0 and 1) between the polygon and the lens are parallel. This is as defined by the property of the lens, which places rays parallel to the axis when directed at the principal plane to and from an axial focal point. The reflected ray 2 is symmetric to the incident ray 1 about the normal n_1 and becomes the ghost ray that enters the lens.

Location of the point of incidence P_2 on the image surface due to ray 2 requires the angle of ray 2 with respect to the axis. From the above data, the two normals n_0 and n_1 are separated by 30° and n_0 is oriented 22.5° below (CW to) the axis. Thus the ghost facet normal n_1 is directed 7.5° above (CCW to) the axis. This places ray 2 at 15° CCW to ray 1. Because ray 1 is parallel to the axis, ray 2 enters the lens at 15° CCW with respect to the axis. This allows application of another property of the principal plane: The refracted ray 2 terminates on the image surface at the same point as that of a ray from the back nodal point P_B propagating at the same 15° angle. That is, point P_2. (Consider ray 2 as part of a very wide collimated beam that fills the lens; all of which rays focus at point P_2, including the one through the lens nodal point P_B. Although a thick lens is given two principal planes, only the back plane need be represented here.)

Having located point P_2 of the ghost ray on the image surface, finding the limit scanned focal point is much simpler. This determines the scan format and the possibility of interference of the ghost within the image. Indicated above was the working field angle of ±15°, identified on Figure 5.7 as (U) for the up-deflected ray and (D) for the down-deflected ray (nomenclature of Fig. 4.2). Applying the same criterion of following the ray from the back nodal point P_B to the image surface along the *same 15° angle* terminates the image format at the *same point* P_2. The input beam angle of 45° is, therefore, borderline.

CHAPTER 6

SUMMARY—MAJOR SCANNER CHARACTERISTICS

This work offers a consolidated presentation of the expanding technology of optical scanning. It expresses relationships between disciplines that are normally considered independently, for pedagogic and operational value. The reader may gain thereby insight to the foundations of the concepts, so that they may be interpreted and used more effectively.

The **Introduction** of Chapter 1 provides a combined view of *active* scanning (as by laser) and *passive* scanning (as for remote sensing). Very similar scanning devices are often used in the reciprocal ray paths of conjugate imaging. This is illustrated by two figures, and two additional figures show operation in both directions. This topic is followed by an introduction to the Lagrange invariant, its relationship to scanned resolution, and the formation of the resolution invariant. This important subject is expanded with a comprehensive coverage in Chapter 3. Chapter 1 closes with an orientation in system architecture, placing the discussion of components and systems into operational perspective. These architectural concepts are extended in several sections of Chapter 4.

The second chapter casts the foundation of the technology with a selection of the fundamentals of **scanning theory**. It presents some unifying analogies between optical and electrical processes that often bear different nomenclature. Factors are highlighted that govern the achievement of the desired image quality when recording or extracting its integrity when scanning. The range of subjects in these two chapters is supported with many original illustrations, rendered to solidify the concepts for later discussion and practical application.

Chapter 3, **Scanned Resolution**, address the remaining building block for successful component and system development, integrating several key items discussed earlier. The basic resolution equation is developed and rendered as a useful nomograph. It is then augmented to provide a more general relationship to account, for example, for a holographic disk scanner that is illuminated with a converging or diverging beam. Additional topics include the aperture shape factor with values identified, the propagation of noise and error components, and the scanned resolution of passive, incoherent, and remote sensing systems.

The first three chapters form the basis for the extensive exposition in Chapter 4, **Scanner Devices and Techniques**. Attention is devoted to its organization with a family tree chart and with discussion oriented to scanner function. Each device is identified within a *high* or a *low* inertia catagory, in performance rather than in sheer physical properties. For example, the galvanometer-type resonant scanner performs as a high-inertia device, although it is of low-inertia structure, and is offered inadvertently as a low-inertia scanner. Included in this chapter, along with the more familiar resource of scanner devices, is the group now emerging from research in the field of *agile beam steering*, represented by the *phased array* and the *decentered microlens array*. A brief review of the devices and techniques covered in this chapter follows.

Initial regard is accorded the (high inertia) rotating polygon for its unique capability for extremely high data transfer rate (bandwidth) and for extremely high resolution at high uniformity and at high efficiency. Special attention is devoted to its design as related to its imaging optics. Polygon use as a device and with optics, some of which were introduced earlier, are developed further. Of similar qualities is the holographic scanner, which may also provide useful correction for cross-scan error. (Chapter 5 addresses the topic of error correction.) Although not providing the high speed of the polygon in its current form, the holographic scanner with matching optics serves most requirements with parallel performance. It requires, however, very astute attention to its design and fabrication.

The topic of *image rotation* is included, in which the focused spot contour (or array of spots) may rotate about its own axis during scanner rotation. This condition explained, the methods of *image derotation* are addressed. Additionally, the topic of *passive scanning* for remote sensing is discussed. A system is presented that both scans the remote field and uses the derived video signal to modulate and the same rotating polygon to provide the scans that record the image, in real time, of the remotely sensed field.

The rotational scanner catagory is followed by the oscillatory devices, the galvanometer and its counterpart, the resonant scanner. Both render valued performance with unique characteristics. The modern galvanometer can provide wide-angle recurrent scanning and effective random-access beam positioning at speed governed by its torque and inertia. The resonant scanner, however, oscillates harmonically. It is capable of high scan rates, albeit with sinusoidal excursions. Distinctions of the two devices are highlighted, contrasting their linearity and duty cycle. Substantive correction processes are incorporated for the resonant device to linearize the scan by triggering the scanned data points and accommodating the increased and varying data rate that results from its sinusoidal oscillation. This catagory of vibrational devices includes one of relatively recent interest: the "fast-steering mirror," mounted to pivot about the center of the mirror and actuated by four pistonlike transducers at its edges to scan *in two dimensions*. Although the scan angle is relatively narrow, mirror size can be large, resolution high, and the speed and random access fast. A device of small mechanical excursion, it forms a contender for *agile beam steering* (covered at the end of this chapter.)

A review of *scanner-lens relationships* follows, with added attention to the architecture of double-pass systems. The rarely addressed topic of aperture relay is discussed, for providing the efficient optical transfer of scan angle from one deflector to another. Lens orientation is then referenced in this work for the control of deflected error components and their propagation through a system.

Low-inertia scanning by bulk devices—the acoustooptic (AO) and electrooptic (EO) scanners—is then presented. The conventional uses of the AO devices are discussed, along with rarely reported significant concepts: the *Scophony* process, the *traveling lens*, and the *chirp deflectors*. The EO device is then discussed to include some practical factors such as their drive voltage and power requirements and the consequence of input beam convergence to avoid encountering its walls during deflection. As with the AO devices, the significance of the selected EO materials is indicated.

The new topic of *agile beam steering* completes Chapter 4 with broad and measured coverage of its background, principles, practice, and prospects, concentrating on the most prominent techniques: the *phased array* and the *decentered microlens array*. Analogous to radar antenna technology, the optical phased array utilizes a series of small optical phase modulators, programmed to shift phase progressively in small groups; each group providing a 2π phase change. The assembled groups form the full aperture width, imparting a deflecting slope to the output wavefront. This technology requires critical device development and complex electrical programming. The decentered microlens array accomplishes the phase change differently. Composed of an array of miniature telescopes, one element of each is capable of minute displacement and is ganged with the rest of the shiftable elements in the array. When shifted, the resulting tilted phase from each telescope is summed in the far field to impart beam steering. This technique requires very precise fabrication, positioning, and uniform control of the lenslets over minute distances. Substantive performance has been achieved for both of the above methods. They both form, effectively, a blazed diffraction grating and benefit from its high diffraction efficiency. This technology is summarized separately at the close of Chapter 4, providing some practical observations.

Chapter 5, **Control of Scanner Beam Misplacement**, discusses correction techniques that are rarely addressed. It covers two forms of beam misplacement. The first, "cross-scan error," causes the defect of minutely displaced scan line groups that is perceived as "banding." A variety of options are presented for control of this important problem. The anamorphic method of such error control is discussed generally, providing for adjustment of the magnitude of the correction to minimize any side effect upon the rest of the system. The second problem generates a "ghost" image that forms an interfering vertical line in the field of an otherwise well-formed scanned image. Ghost image correction is also presented with alternates. In one case, possible interaction with the above cross-scan error reduction is avoided. This more general method is detailed, concluding with an expression for the correction that uses the characteristics of and at the polygon, independently of subsequent imaging optics.

6.1 COMPARISON OF MAJOR SCANNER TYPES

Although the devices and their associated techniques are covered intently and referenced appropriately in Chapter 4, little is presented in the literature of their relative virtues and limitations as scanner options. Thus the accompanying Table 6.1 provides brief notations of the **Characteristics of the Major Scanner Types** to serve as a comparison guide. The following discussion expands on its condensed format.

TABLE 6.1 Characteristics of Major Scanner Types

Type	Random Access	Scan Angle	Resolution Elements Per Scan	Scan Rate Scans per Second	Accuracy and Linearity	Cost of Scanner*	Comments
Monogon Fig. 1.10	No	Very wide (to 360°)	>50,000 (intrnl. drum)	Per shaft speed	Very high	Low	Limited duty cycle
Polygon Figs. 3.6 and 3.7	No	Wide per no. of facets	>20,000; <50,000 overfilled	>30,000	Very high; critical repeatabillity	Moderate to low	Very high performance
Holographic Figs. 4.12 and 4.16	No	Wide, per no. of facets	>15,000; <50,000, overfilled	>5000 Limited by substrate	Very high; potentially less critical	Moderate to low	λ-Sensitive; Maturing technology
Galvanometer Figs. 1.8 and 4.19a	Yes	Moderate; limited angle-speed	>10,000	<1000	Very high with feedback	Moderate to low	Associated electronics; available open loop
Resonant Figs. 1.8 and 4.19b	No	Good; limited angle-speed	<10,000	>15,000, in both scan directions	High; electronic compensation	Moderate to low	Associated electronics; limited duty cycle
Acoustooptic Fig. 4.22	Yes	Narrow (expand optically)	<2000	<20,000	Per electronics	Moderate	Associated electronics and optics
Electrooptic Fig. 4.28	Yes	Narrow (expand optically)	<100	>50,000; (extremely fast)	Per electronics	Moderate	Associated electronics

*Cost is strong function of deflecting aperture size, speed, accuracy and repeatability.

Most prominent is the *rotating polygon* for its effectiveness and utility, such as selectable scan angles, high scan rates, high resolution, and high accuracy and linearity. Laser printers, for example, employ rotating prismatic polygons. Some graphic arts imagesetters employ the monogon (and its stabilized variations) and the prismatic and pyramidal polygons. (Table 4.1, compares *prismatic* and *pyramidal* polygons.) The pyramidal polygon is utilized less often, limited typically by its higher fabrication cost. Where, however, optical packaging is advantageous, or where radial symmetry is useful, the pyramidal polygon provides similar performance. Both can provide resolutions to—and beyond—20,000 elements/scan at data rates in the 100-mHz range. The simultaneous combination of such high resolution and speed requires substantial discipline in design and fabrication.

The *holographic* scanner runs quite parallel to the polygon in utility and systematic performance, offering advantages in Bragg angle wobble correction and in low aerodynamic loading. Thus, when properly designed, the holographic scanner can outperform the polygon in speed. Current technology utilizing transparent (often glass) disk substrates is limited by the inertial burden on these materials as a function of increased speed. In general, its critical technology demands high expertise in development and production, constraining the availability of these devices. This can change when the holographic scanner becomes fabricated in quantity by replication on a stable and uniform substrate, while retaining the detailed requirements of accuracy and the integrity of its grating contour.

Because of its operational adaptability, the *galvanometer* is often selected for experiment and application to low-speed recurrent scanning or to precise random access positioning. The modern galvanometer with feedback is capable of accuracy that challenges that of the high inertia rotational scanners. At speeds to several hundred hertz, the scan angle becomes limiting. The *resonant scanner* is capable of much higher useful scan rates, forming sinusoidal oscillations at a fixed frequency. Although the product of speed and resolution can be high, resonant devices require adaptations for electronic pixel linearization and increased exposure energy because of nonuniform velocity and restricted duty cycle, especially if scan is one-directional.

The *acoustooptic* (AO) scanner is the most popular "nonmechanical" modulator and deflector. It is capable of extremely high scan rates, to 20,000 per second. As scan rate is increased, however, its resolution and duty cycle are reduced because of its fixed "retrace" interval. Thus resolution rarely exceeds 1000 elements per scan. The AO device can serve also as a rapid "random access" device, limited by the time needed to fill the aperture with the new acoustic wave. Unfortunately, given a fixed Bragg angle (per input beam angle and drive frequency), a changing drive frequency, which provides scan, also makes the acoustic wave shift the optical Bragg regime. This is accompanied by a loss in diffraction efficiency, which may be controlled to within a 10% variation at the margins of scan. Throughput efficiencies range between 50% and 90%.

The *electrooptic* (EO) scanner, another nonmechanical device, is capable of the highest speed at, however, lower resolution. Requiring only the time to charge and discharge a small value of capacitance, the speed of the EO deflector is limited mainly by the power dissipation (loss tangent) in its electrooptic materials and the voltage gradients at the drive waveforms and frequencies.

Alternate scanner techniques not explicit in this work include piezo-electric (mirror) beam deflection, digital deflection (binary EO switching), intracavity scanning (mode selection within a laser resonator), beam iteration (angle transfer), and scan amplification (near-resonant iteration for increased resolution). All the above are discussed elsewhere [Bei2]. Additionally, rotating prism scanners have been analyzed [Wol] and such recent broadband work has been detailed [Lac]. Related to intracavity scanning, work in electron beam-pumped semiconductor laser radiation sites is of interest [Nas].

REFERENCES

Adl   R. Adler, "Interaction between light and sound", *IEEE Spectrum*, 4, (5), 42–54 (May 1967).

Alp   L.J. Alpet, et al., "Laser array techniques", Quart. Report 3A, Air Force Contract AF33(615)3918, AD No. 821235 (1967).

Ayl   R.P. Aylward, "Advances & technologies of galvanometer-based optical scanners", in *Optical Scanning: Design & Applications*, Proc. SPIE Vol. 3787, L. Beiser, S. Sagan & G. Marshall, Eds., pp. 158–164 (1999).

B&J   L. Beiser and R.B. Johnson, "Scanners", in OSA *Handbook of Optics*, Vol. 2, M. Bass, Ed., Chapt. 19, McGraw Hill, New York (1995).

B&S   E. Barkan and J. Swartz, "Depth of Modulation and Spot Size Selection in Bar Code Laser Scanners. *Proc. SPIE* Vol. 299, L. Bciser Ed., pp. 82–97 (1981). Also, *SPIE Milestone Series: Laser Scanning and Recording*, Vol. 378, L. Beiser, Ed., pp. 66–80 (1985).

B&Y   W.H. Bliss and C.J. Young, "Facsimile scanning by a cathode ray tube", *RCA Review*, XV, No. 3, 275–290 (Sept. 1954).

Bade   L. Bademian, "Acousto-optic laser recording", *Opt. Engr.*, Vol. 20, No. 1 143–149 (Jan/Feb. 1981).

Ball   *Fast Steering Mirrors*, data & specifications, Ball Aerospace & Technologies Corp., Boulder, CO 80301.

Bar   R.B. Barnes, U.S. Pat, No. 3,287,557 (1996).

Bei1   L. Beiser, *Holographic Scanning*, John Wiley & Sons, New York (1988).

Bei2   L. Beiser, "Laser Scanning Systems", in *Laser Applications*, Vol. 2, M. Ross, Ed., Academic Press, New York, pp. 55–159 (1974).

Bei3   L. Beiser, "Fundamental Architecture of Optical Scanning Systems", *Appl. Opt.* Vol. 34, No. 31, pp. 7307–17 (1 Nov. 1995).

Bei4   L. Beiser, "Generalized Equations for the Resolution of Laser Scanners", *Appl. Opt.*, Vol. 22, No. 20, 3149–51 (15 Oct. 1983).

Bei5   L. Beiser, "Design equations for a polygon laser scanner", in *Beam Deflection and Scanning Techniques*", Proc. SPIE Vol. 1454, G.F. Marshall & L. Beiser, Eds., pp. 60–66 (1991).

Bei6   L. Beiser, "Spot distortion during gradient deflection of focused laser beams", *IEEE J. Quant. Electr.*, Vol. QE-3, No. 11, 560–67 (Nov. 1967).

Bei7   L. Beiser, "Optical Scanners", in *Encyclopedia of Applied Physics*, Vol. 12, pp. 337–368, VCH Publishers, Inc., NY (1995).

Bei8   L. Beiser, *Laser Scanning Notebook*, SPIE Press, Bellingham, Wa (1992).

Bei9   L. Beiser, U.S. Pat. Nos. 4,963,643 (1990) and 5,114,217 (1992).

Ber   A. Berta, "Development of high performance fast steering mirrors", in *Optical Scanning: Design and Applications*, Proc. SPIE, Vol. 3787, L. Beiser, S. Sagan & G. Marshall, Eds., pp. 181–192 (1999).

Boyd   R.W. Boyd, *Radiometry and the Deflection of Optical Radiation*, pp. 89–93 John Wiley & Sons (1983).

Bre   M.S. Brennesholtz, "Light collection efficiency for light valve projection systems", *Proc. SPIE* Vol. 2650, M.H. Wu, Ed., pp. 71–79 (1995/6).

Bro   E.B. Brown, *Modern Optics*, Reinhold Publishing Corp., N.Y. (1965).

BTL   Bell Telephone Laboratories, Members of Tech. Staff, *Transmission Systems for Communications*, Bell Telephone Laboratories, Inc. (Dec. 1971).

Bur   R.S. Burington, *Handbook of Mathematical Tables and Formulas*, 2nd Ed., p. 29, Handbook Publishers, Inc., Sandusky, Ohio (1940).

Burn   D.M. Burns, V.M. Bright, S.C. Gustafson and E.A. Watson, "Optical beam steering using surface machined gratings and optical phased arrays". *Optical Scanning Systems: Design and Applications*, SPIE Vol. 3131, L. Beiser and S. Sagan, Eds., pp. 99–110 (1997).

C&F   T.C. Cheston and J. Frank, *Radar Handbook*, M.I. Skolnick, Ed., McGraw Hill, NY (1970).

C&R   B.R. Clay & D.M. Rowe, U.S. Pat. No. 5,182,659 (1993).

CBS   Private Communications, R.H. McMann, CBS Labs., Stamford, CT. (1967).

CG&A   G.A. Coquin, J.P. Griffin and L.K. Anderson, "Wideband acoustooptic deflectors using acoustic beam steering", *IEEE Trans. Sonics & Ultrasonics*, SU-17(1), pp. 34–40 (Jan. 1970).

D&J   L.C. DeBenedictis & R.V. Johnson, U.S. Pat. No. 4,205,348, "Laser Scanning utilizing facet tracking & acoustic pulse imaging techniques" (May 1980).

D&S   J.C. Dainty and R. Shaw, *Image Science*, Academic Press, N.Y. (1974).

D&Se   P. Debye and F.W. Sears, "On the scattering of light by supersonic waves", *Proc. Nat'l. Acad. Sci.* (USA) 18, 409 (1932). Also, R. Lucas and P. Biquard, "Optical properties of solids and liquids under ultrasonic vibrations", *J. Phys. Rad.*, 7th Ser., 3, 464 (1932).

Don   J.P. Donahue, "Laser pattern generator for printed circuit board artwork generation", SPIE Milestone Series: *Laser Scanning and Recording*, Vol. 378, L. Beiser, Ed., pp. 421–428 (1985).

Dor   T.A. Dorschner, R.C. Sharp, D.P. Resler, L.J. Friedman & D.C. Hobbs, "Basic laser beam agility techniques", Wright Patterson Air Force Base, WL-TR-93-1020 (AD-B-175-883) (1993).

Dro   W.S. Drozdoski, Booz-Allen & Hamilton Inc., private communication.

FC&C   L.C. Foster, C.B. Crumly & R.L. Cohoon, "A high resolution linear optical scanner using a traveling wave acoustic lens", *Appl. Opt.*, Vol. 9, No. 9, (Sept. 1970).

F&S   V.J. Fowler and J. Schlafer, *Appl. Opt.*, Vol. 5, No. 10, p. 1675 (1966).

Farn   M.W. Farn, "Agile beam steering using phased array-like binary optics", *Appl. Opt.*, Vol. 33, No. 22, 5151–5158 (Aug. 1994).

Fau   M. Fauver, et al, "Micro fabrication of fiber-optic scanners," Phoc. SPIE Vol. 4773, *Optical Scanning 2002*, S. Sagan, G. Marshall & L. Beiser, Eds., pp. 102–110 (2002).

Fle   J.M. Fleischer, "Light scanning and printing systems", U.S. Pat. No. 3,750,189 (1973). Also, J.M. Fleischer, M.R. Latta & M.E. Achedrau, "Laser optical system of the IBM 3800 printer", *IBM J. Res. Dev.*, Vol. 21, p. 480 (1977).

Flo   K.M. Flood, B. Cassarly, C. Sigg & J.M. Finlan, "Continuous wide angle beam steering using translation of binary microlens arrays and a liquid crystal phased array", in *Computer and Optically Formed Holographic Optics*, Proc SPIE Vol. 1211, pp. 296–304 (1990).

Fow   V.J. Fowler, et al., "Electrooptical light beam deflection", *Proc. IEEE*, Vol. 52, p. 193 (1964).

G&B   A.J. Giarolla and T.R. Billeter, "Electrooptic deflection of a coherent light beam", *Proc IEEE*, Vol. 51, p. 1150 (1963).

G&H   W. Goltsos and M. Holz, "Agile beam steering using binary optics microlens arrays, *Opt. Engr.*, Vol. 29, No. 11, pp. 1392–1397 (Nov. 1990).

G&M   J.S. Gerig and H. Montague, *Proc. IEEE* Vol. 52, p. 1753 (1964).

Gad   J.S. Gadhok, "Achieving high duty cycle sawtooth scanning with galvanometric scanners", in *Optical Scanning: Design and Application*, SPIE Vol. 3787, L. Beiser, S. Sagan & G. Marshall, Eds., pp. 173–180 (1999).

Gin   R.H. Ginsberg, "Image Rotation", Engr. & Lab. Notes, *Optics & Photonics News*, Vol. 6, No. 2, 5–8 (Feb. 1994).

Good   D.S. Goodman, "General Principles of Geometrical Optics", in OSA *Handbook of Optics*, Vol. 1, M. Bass, Ed., Chapt. 1, McGraw Hill, New York (1995).

Gor   E.I. Gordon, "A review of acoustooptical deflection and modulation devices" *Proc. IEEE*, Vol. 54 (10), 1391–1401 (1966).

Got   M. Gottlieb, "Acoustooptic scanners and modulators", in *Optical Scanning*, Ed. G.F. Marshall, pp. 615–685, Markel Dekker, Inc., NY (1991).

Gra   P.G. Grant, R.A. Meyer and D.N. Qualkinbush, "An optical phased array beam steering technique", *Proc. Electro-Opt. Syst. Des. Conf.* pp. 259–264 (1971).

Hae   G.H. Haertling, "PLZT electrooptic materials and applications—a review", *Ferroelectrics*, Vol. 75, pp. 25–55 (1987).

Hob   P.C.D. Hobbs, *Building Electro-Optical Systems*, John Wiley & Sons, Inc. (2000).

H&Bo   O. Hadar and G.D. Boreman, "Oversampling requirements for pixelated-imager systems", *Opt. Engr.*, Vol. 38, No. 5, 782–785 (May 1999).

H&Bu   R.E. Hopkins and M.J. Buzowa, "Optics for laser scanning", in SPIE Milestone Series: *Laser Scanning and Recording*, Vol. 378, L. Beiser, Ed., 123–127 (1985). Also in *Opt. Engr.*, Vol. 15, No. 2, 90–94 (April 1976).

H&S   R.E. Hopkins & D. Stephenson, "Optical systems for laser scanning", in *Optical Scanning*, pp. 27–81, Ed. G.F. Marshall, Markel Dekker, Inc., NY (1991).

I&L   C.L. Ireland and J.M. Ley, "Electrooptical Scanners", in *Optical Scanning*, pp. 687–778, Ed. G.F. Marshall, Markel Dekker, Inc., NY (1991).

Ike   H. Ikeda, et al., "Hologram scanner", *Fujitsu Sci. Tech. J.*, 23:3 (1987).

J&M   R.H. Johnson & R.M. Montgomery, "Optical beam deflection using acoustic-traveling-wave technology", *Proc. SPIE*, Acousto-Optics/Instrumentation/Applications, Vol. 90, p. 43 (Aug. 1976).

J&W   F.A. Jenkins and H.G. White, *Fundamentals of Optics*, McGraw Hill, NY (1957).

Joh   R.V. Johnson, "Scophony Light Valve", *Appl. Opt.*, 18 (23) pp. 4030–38 (Dec. 1979).

John   R.B. Johnson, "Radar analogies for optics", Proc. SPIE, Vol. 128, *Effective Utilization of Optics in Radar Systems*, Ed.: B.W. Vatz, pp. 75–83 (1977).

JG&S   R.V. Johnson, J. Geurin and M. Swanberg, "Scophony spatial light modulator" *Opt. Engr.* Vol. 24, No. 1, 93–100 (1985).

Kay   D.B. Kay, "Optical Scanning System with Wavelength Correction", U.S. Pat. No. 4,428,643 (1984).

Kell   R.D. Kell, A.V. Bedford and M.A. Trainer, "Scanning Sequence and Repetition Rates of Television Images", *Proc. IRE*, Vol. 22, p. 1246 (1934).

Kes   D. Kessler, D. DeJager & M. Noethen, "High resolution laser writer", in *Hard Copy Output*, Proc. SPIE Vol. 1079, L. Beiser, Ed., pp. 27–35 (1989).

Keys   A.S. Keys, R.L. Fork, T.R. Nelson, Jr. & J.P. Loehr, "Resonant transmissive modulator construction for use in beam steering array", SPIE Vol. 3787, *Optical Scanning: Design & Application*, pp. 115–125, L. Beiser, S. Sagan & G.F. Marshall, Eds. (July 1999).

Kle   J.E. Klein, "Geometrical relationships characterizing polygonal scan wheels, Proc. SPIE Vol. 554, D.T. More & W. Taylor, Eds., pp. 468–477 (1985).

Kor   A. Korpel, "Acousto-optics", in *"Appl. Opt. & Opt. Engr."*, Vol. VI, R. Kingslake & B.J. Thompson, Eds., pp. 89–141 Academic Press, NY (1980).

Kra1   C.J. Kramer, U.S. Pat. No. 4,289,371, "Optical scanner using plane linear diffraction gratings on a rotating spinner" (Sept. 1981).

Kra2   C.J. Kramer, U.S. Pat. No. 4,239,326, "Holographic scanner for reconstructing a scanning light spot insensitive to mechanical wobble. (Dec. 16, 1980).

Kra3   C.J. Kramer, Holographic deflector for graphic arts systems", in *Optical Scanning*, pp. 213–349, Ed. G.F. Marshall, Marcel Dekker, Inc., NY (1991).

L&B   K.S. Lee & F.S. Barnes, "Microlenses on end of optical fibers for laser applications", *Appl. Opt.*, Vol. 24, No. 19, pp. 3134–3139 (1985).

L&K   Y. Li & J. Katz, "Asymmetric distribution of the scanned field of a rotating reflective polygon", *Appl Opt.*, Vol. 36, No. 1, pp. 342–352 (1 Jan. 1997).

L&U   D.R. Lehmbeck and J.C. Urbach, "Scanned Image Quality", in *Optical Scanning*, Ed. G.F. Marshall, pp. 88–91, Marcel Dekker, Inc., NY (1991).

L&Z   T.C. Le and J.D. Zook, "Light beam deflection with electrooptic prisms", *IEEE J. Quant. Electr.*, Vol. QE-4, No. 7, pp. 442–454 (1968).

Lac J. Lacoursiere, et al., "Large deviaton achromatic Risley prisms pointing systems", in *Optical Scanning II*, Proc. SPIE Vol. 4773, S. Sagan, G.F. Marshall & L. Beiser, Eds., (July 2002).

Len B.A. Lengyel, *Lasers*, 2nd Ed., John Wiley & Sons., N.Y. (1971).

Lenn E. Lennemann, "Aerodynamic aspects of disk files", IBM J. Res. Dev. 480–488 (Nov. 1974).

Lev L. Levi, *Applied Optics*, Vol. 1, John Wiley and Sons, New York (1968).

Lew N.W. Lewis, "Television Bandwidth and the Kell Factor", *Electronic Technology* (England), pp. 44–47 (Feb. 1962).

Los J.F. Lospeich, "Electro-optic light beam deflectors", *IEEE Spectrum*, Vol. 5, No. 2, pp. 45–52 (Feb. 1968).

M&G G.F. Marshall & J.S. Gadhok, "Resonant and galvanometric scanners: integral deflector position sensing", *Photonics Spectra*, 155–160 (June 1991).

Mc&W P.F. McManamon and E.A. Watson, "Optical beam steering using phased array technology", SPIE Vol. 3131, *Optical Scanning Systems: Design and Applications*, pp. 90–98, L. Beiser & S. Sagan, Eds., (July 1997).

M&W T.D. Milster and J.N. Wang, "Modeling and measurement of a micro-optic beam deflector", in *Design, Modeling and Control of Laser Beam Optics*, Y. Kohanzadeh, G.W. Lawrence, J.G. McCoy and H. Weichel, Eds., Proc. SPIE Vol. 1625, pp. 78–83 (1992).

M&Z G.M. Morris and D.A. Zweig, "White light Fourier transforms", in *Optical Signal Processing*, Chapt. 1.2, J.J. Hovner, Ed., Academic Press, NY (1987).

Mac A.J. MacGovern, U.S. Pat. No. 3,910,675, "Laser scanner apparatus", (Oct. 7, 1975).

Mara A.S. Marathey, "Diffraction" in *Handboodk of Optics*, Vol. 1, M. Bass, Ed., Chapt. 3. (Optical Society of America) McGraw Hill, NY (1994).

Mar1 G.F. Marshall, "Center-of-scan locus of an oscillating or rotating mirror, in *Laser Scanning and Recording Systems*, Proc. SPIE Vol. 1987, L. Beiser, Ed., pp. 221–231 (1993).

Mar2 G.F. Marshall, T.S. Vettese & J.H. Caroselia, "Butterfly line scanner", in *Beam Deflection and Scanning Technologies*, SPIE Proc. Vol. 1454, G.F. Marshall & L. Beiser, Eds., pp. 37–45 (1991).

Mar3 G.F. Marshall, "Geometrical determination of the positional relationship between the incident beam, the scan axis and the rotation axis of a prismatic polygon scanner", in *Optical Scanning 2002*, Proc. SPIE Vol. 4773, S.F. Sagan, G.F. Marshall & L. Beiser, Eds., pp. 38–51 (July 2002).

Mar4   G.F. Marshall, "Stationary ghost images outside the image format of the scanned field image plane", in *Optical Scanning 2002*, Proc. SPIE Vol. 4773, S.F. Sagan, G.F. Marshall & L. Beiser, Eds., pp. 132–140 (July 2002).

Mas   K. Masaki, et al., U.S. Pat. No. 4,040,737 (Aug. 1977).

Mat   Matsumoto, et al., U.S. Pat. No. 4,176,907 (1979).

McD   G.F. McDearmon, K.M. Flood, and J.M. Finlan, "Comparison of conventional and microlens-array agile beam steerers", Proc. SPIE Vol. 2383, *Micro-Optics/Micromechanics and Laser Scanning and Shaping*, M.E. Motamedi & L. Beiser, Eds., pp. 167–178 (Feb. 1995).

McM   P.F. McManamon, et al., "Optical Phased Array Technology", *Proc. IEEE*, Vol. 84, No. 2, pp. 268–298 (Feb. 1966).

Mey   R.A. Meyer, "Optical beam steering using a multichannel lithium tantalate crystal", *Appl. Opt.*, Vol. 11, No. 3, p. 613 (March 1972).

Min   K. Minoura, U.S. Pat. Nos. 4,993,792 (Feb. 1991) & 5,191,463 (March 1993).

Mon1   J. Montagu, "Galvanometric and resonant low inertia scanners", in *Optical Scanning*, G.F. Marshall, Ed., Marcel Dekker Inc, NY, 525–613 (1991).

Mon2   J. Montagu, "Novel scanner designs", in *Optical Scanning 2002*, Proc. SPIE Vol. 4773, S.F. Sagan, G.F. Marshall & L. Beiser, Eds., pp. 1–10 (July 2002).

Mur   H. Muramatsu, et al., "Dynamic etching method for fabricating a variety of tip shapes in the optical fiber probe of a scanning near-field microscope", *J. Microscopy*, Vol. 194, pt2/3, pp. 383–387 (1999).

Nas   A.S. Nasibov, et al., *J. Crystal Growth*, Vol. 117, pp. 1040–1045 (1992).

New   *Fast Steering Mirrors* data & specifications, Newport Corp, Irvine, CA 92606.

Osh   D.C. O'Shea, *Elements of Modern Optical Design*, John Wiley & Sons (1995).

P&T   G.B. Parent & B.J. Thompson, *Physical Optics Notebook*, SPIE (1969). Also, with G.O. Reynolds & J.B. DeVelis, *The New Physical Optics Notebook: Tutorials in Fourier Optics*, SPIE Vol. PM01 (1989).

Poh   K.C. Pohlmann, *The Compact Disc Handbook*, 2nd Ed, A-R Editions, Inc., Madison, Wisconsin (1992).

Pro   L.W. Procopio, F.A. Jessen and L.J. Brown, "Laser phased arrays", Proc. *8th International Conf., Mil. Electron. Groups*, Contract AF30(602)-2901 (1964).

R&M   J. Randolph and J. Morrison, "Modulation transfer characteristics of an acoustooptic deflector", *Appl Opt.*, Vol. 10, No. 6, 1383–85 (June 1971).

Rei   S. Reich, "Use of electromechanical mirror scanning devices", *SPIE Milestone Series: Laser Scanning and Recording*, Vol. 378, L. Beiser, Ed., pp. 229–238 (1985).

Row   D.M. Row, "Developments in holographic-based scanner design", in *Optical Scanning Systems: Design and Applications*, Proc. SPIE Vol. 3131, L. Beiser and S. Sagan, Eds., pp. 52–58 (1997).

S&T   B.E.A. Saleh & M.C. Teich, *Fundamentals of Photonics*, John Wiley & Sons, Inc. (1991).

Sch   W.F. Schreiber, "Laser dry silver recorder", in *SPIE Milestone Series: Laser Scanning and Recording*, Proc. SPIE, Vol. 378, L. Beiser, Ed., pp. 437–441 (1985). Also in SPIE Vol. 53, *Laser Recording and Informtion Handling*, L. Beiser, Ed., pp. 116–121 (Aug. 1974).

Sei   E.J. Seibel, et al., "Single fiber flexible endoscope: general design for small size, high resolution and wide field of view", Proc. SPIE Vol. 4159, *Biomonitoring and Endoscopy Technologies*, pp. 29–39 (2001).

She   R.J. Sherman, "Polygonal scanners: application, performance and design", in *Optical Scanning*, G.F. Marshall, Ed., Marcel Dekker Inc., NY pp. 351–450 (1991).

Sherr   S. Sherr, *Applications for Electronic Displays*, John Wiley & Sons, Inc., NY (1998).

Shep   J. Shepherd, "Windage of rotating polygons" (Ref. same as She above) pp. 451–476.

Shu   A.R. Shulman, *Optical Data Processing*, John Wiley & Sons, Inc., NY (1970).

Sko   M.I. Skolnik, *Introduction to Radar Systems*, McGraw Hill, NY (1962).

Sta   G.S. Starkweather, U.S. Pat. No. 4,475,787 (1984).

Sto   J.E. Stockley, S.A. Serati, G.D. Sharp, P. Wang and K.M. Johnson, "Broadband beam steering", *Optical Scanning Systems: Design and Applicatons*, Proc. SPIE, Vol. 3131, L. Beiser & S. Sagan Eds., pp. 111–123 (1997).

Str   R.J. Straayer, U.S. Pat. No. 4,978,184, "Laser raster scanner having positive facet tracking", (Dec. 18, 1990).

Swa   G.J. Swanson, "Binary optics technology: theory and design of multi-level diffractive optics elements", *Lincoln Labs Tech. Rep. 854* (Aug. 14, 1989).

Swe   M. Sweeney, G. Rynkowski, M. Ketabchi & R. Crowley, "Design considertions for fast steering mirrors (FSMs)", in *Optical Scanning 2002*, Proc. SPIE Vol. 4773, S. Sagan, G.F. Marshall & L. Beiser, Eds., pp. 63–73 (July 2002).

T&K   C.S. Tsai and J.M. Kraushaar, Proc. Electro-Opt. Syst. Des. Conf., p. 176 (1972).

Tho   J.A. Thomas, M. Lasher, Y. Fainman and P. Sultan, "A PLZT-based
           dynamic diffractive optical element for high speed, random access
           beam steering", in *Optical Scanning Systems: Design & Applica-
           tions*, Proc. SPIE, Vol. 3131, L. Beiser & S. Sagan, Eds., pp. 124–132
           (1997).

Toy   G. Toyen, "Generation of precision pixel clock in laser printers and
           scanners" in SPIE Milestone Series, *Laser Scanning & Recording*,
           Vol. 378, L. Beiser, Ed., pp. 315–322 (1985). Also in *Laser Scanning
           Components and Techniques*, Proc. SPIE Vol. 84, L. Beiser &
           G. Marshall, Eds., pp. 138–145 (1976).

Tuc   I. Tuchman, "Laser scanning and chopping methods using mechan-
           ical resonant devices", *Proc. SPIE*, Vol. 3787, L. Beiser & G.
           Marshall, Eds., pp. 165–172 (1999).

Twe   D.G. Tweed, "Resonant scanner Linearization techniques", *Opt.
           Engr.* Vol. 24, No. 6, 1018–1022 (1985).

W&M   E.A. Watson and A.R. Miller, "Analysis of optical beam steering
           using phased micromirror arrays", *Proc. SPIE*, Vol. 2687, pp. 60–67
           (1996).

Wal   C.T. Walters, "Flat field postobjective polygon scanner", *Appl. Opt.*,
           Vol. 34, No. 13, pp. 2220–25 (1 May 1995).

Wat1   E.A. Watson, "Analysis of beam steering with decentered microlens
           arrays", *Opt. Engr.*, Vol. 32, No. 11, pp. 2665–2670 (Nov. 1993).

Wat2   E.A. Watson, D.T. Miller & P.F. McManamon, "Applications and
           requirements for nonmechanical beam steering in active electro-
           optic modulators", in *Diffractive and Holographic Technologies,
           Systems and Spatial Light Modulators VI*, I. Cindrich, S.H. Lee &
           R.L. Sutherland, Eds., Proc. SPIE Vol. 3633, pp. 216–225 (1999).

Wat3   E.A. Watson, P.R. McManamon, L.J. Barnes and A.J. Carney,
           "Applications of dynamic gratings to broad spectral band beam
           steering", SPIE Vol. 2120, *Laser Beam Propagation and Control*
           (1994).

Wat4   E.A. Watson, W.E. Whitaker, C.D. Brewer & S.R. Harris, "Imple-
           menting optical phased array beam steering with cascaded
           microlens arrays", *Proc. IEEE Aerospace Conference*, paper 5.036
           Big Sky, MT (March 2002).

Wol   W.L. Wolfe, "Optical mechanical scanning techniques, Chapt 10 in
           The Infrared Handbook, W.L. Wolfe and G.J. Zissis, Eds., ERIM
           (Environmental Research Institute of Michigan) (1978).

YW&M   S.K. Yao, D. Weid and R.M. Montgomery, "Guided acoustic travel-
           ing wave lens for high-speed optical scanners", *Appl. Opt.*, Vol. 18,
           No. 4, (15 Feb. 1979).

Yam   F. Yamagishi, et al., "Lensless holographic line scanner", *Proc. SPIE*
           Vol. 615, 128–132 (1986).

Z&L   J.D. Zook and T.C. Lee, *Proc. SPIE*, Vol. 14, p. 281 (1970).

INDEX

Aberration 24, 42, 70, 82
  off-axis 77
Acoustooptic scanner, *see* Scanning
  devices
Aerodynamic effect 42
Agile beam steering, *see also*
  Scanning devices xii, 3, 105,
  128–145, 162
Airy disc 21, 22, 77
Airborne line scanning 61, 62
Amplitude/intensity functions 19–25
Analogies xii, 2, 3, 19, 20, 27, 87,
  90–92, 130, 131
Anamorphic optics 10, 112, 150–152
Aperture
  clear 23
  Gaussian illuminated 52, 53
  keystone 51–53, 68
  options 68, 69
  overfilled 68, 69, 74–81
  round; elliptic 21, 22, 50, 51, 68
  shape factor 22–24, 47, 50–54, 162
  slit 21
  square; rectangular 20–22, 50, 68
  triangular 51–53, 68
  underfilled 68, 69, 74–81

uniformly illuminated 20–22,
  50–53
Architecture 3, 12–18, 108–112, 161
  aperture relaying 111, 112
  double-pass 74, 75, 109–111
  lens relationships 112
  objective scan 4, 13, 16, 17
  postobjective scan 4, 7, 15, 16, 108
  preobjective scan 4, 13–15, 108
  pupil relief distance 8, 14, 69–73,
  84, 108, 155
Ax-blade scanner 66, 152

Bandwidth, signal 10, 70, 162
Bar code 1
Beam divergence 11
Beam expander/compressor 11, 121
Beam misplacement 162
  along-scan error 54–56, 66, 147,
  148
  banding 147, 148
  cross-scan error 54–56, 66,
  147–155, 164
  wobble 70, 147
  polygon nonuniformities 147
  shaft precison 148

Beam misplacement correction 162
  active, passive 149
  anamorphic 150–152
    original description 151
    quadrature resolution 150
    telescopic compression 151
    wedge beam 151
    zero beam height 151
  Bragg condition 118, 119
  Bragg, holgaphic 88–92, 149
  complementary phase 152
  double reflection 151–155
    butterfly scanner 152, 153
    open mirror scanners 152, 153
    pentaprism-mirror scanners 152, 153
    porro prism (external) 153, 154
  fabrication accuracy 149, 150
  start-of-scan (S.O.S.) 148
  synchronous pilot beam 148
Bearing noise 55
Beryllium substrate 93
Bessel function 21
Bow-tie distortion 62

Cathode ray tube (CRT) 27
Chirp deflector, *see* Scanning devices
Classification of scanners 63, 102
Collimation 5, 8, 32
Compact disk 1
Comparison of scanners, *see* Scanning devices
Conjugate image, *see also* Image 4–9
Convolution function 26, 27

Depth of field; of focus, *see* Scanning, quality
Design considerations, *see* Scanning devices
Diffraction
  Bragg 88, 113–115
  efficiency 115, 143
  ghost spots 15
  limit 11
Diffractive Optics 85, 95

blazed grating 139, 142–144, 163
surface relief grating 114
volume grating 114, 115
Duty cycle, *see* Scanning

Electromagnetic field
  amplitude 19
  intensity 19
Electrooptic, *see also* Scanning devices
  deflection 124–128
  materials 126, 128, 137–139
  modulation 138
  phased array 144
Error components, angular, *see also* Beam misplacement 54–56, 66
Error propagation 54–56, 162
Étendue 12

Fast steering mirror, *see* Scanning devices, Oscillatory
Fiber optic scanner, *see* Scanning devices
Field of view 62
Fill factor 142, 143
Flux density 20
Flux lines 20
F-number of beam 33–36
Flying spot, *see* Scanning
Focus quality, *see* Scanning
Focal spot, *see* Point spread function
Fourier function 25, 133
Frequency, signal 19

Galvanometer, *see* Scanning devices, Oscillatory
Gaussian beam
  development of 31–34
  focus characteristics 35–37
  focus, depth of 36, 37
  propagation 31–34
  waist 31–33
Gaussian (intensity) function 22–24
  $1/e^2$ value 22, 23, 108
  $1/e^2$ to FWHM relationship 24

clear aperture 23
FWHM 23, 24
Rayleigh range 32–34
untruncated 24
Ghost image elimination 147, 155–159, 164
  beam offset method 156
    allows anamorphic correction 156
    determined at polygon 157, 159, 164
    with flat-field lens 157, 159
  fixed reimaged scatter 147, 155, 158
  skew beam method
    anamorphic correction conflict 156, 164
    incidental scan bow 156
    schlieren technique 156
Grating, *see also* Diffractive optics
  blazed 133
  equation 116, 123, 132

Holographic scanner. *see* Scanning devices

Image
  conjugates 4–9, 83, 151, 161
  derotation 82, 83
  error reduction, *see* Invariant
    rotation 18, 80–82, 162
  space 5, 15, 18, 83
Impulse response 20
Information
  conversion 1
  hybrid, parallel, serial xi
  reciprocal 1–3, 19, 83
  transfer 1
Input system 8, 9
Invariant
  Noise (error) reduction 54–56, 82
  Optical 9–11, 54, 161
  Resolution 9–12

Kell factor 31, [Kell]
Korpel, A. 115, [Kor]
Kramer, C. 88, 89, 95–97, 99, [Kra]

Lagrange invariant 9–11, 54, 161
Laser xii, 22, 137
Laser scanning
  definition xii
  incoherent analysis for MTF 42, 43
  tandem 137, 138
Lens
  field 140–142
  flat field (f · Θ, "scan lens") 7, 8, 14, 15, 47, 56, 70–73, 150–154
  objective 4, 7, 15
  on axis 16
  pupil relief distance, *see* Architecture
Light
  coherence 117, 118
  modulation, *see also* Modulation 84
Line spread function, *see also* Point spread function 2

Meridional plane 10
Mirror
  apertured 6
  turning 6
  partially silvered 6
Modulation 9, 26, 39, 66, 117–119
Modulation transfer function (MTF) 37–45, 52, 53
  aperture covolution; autocorrelation 40, 41, 52
  cascaded (tandem) stages 42
  coherent effects 119
  Fourier series 40
  Fourier transform 38–40
  incoherent MTF 42, 118
  modulation depth 39
  monotonic decrement; typical 43
  MTF = 0.5 quality descriptor 43
Monogon, *see* Scanning devices

Near field; far field 20, 23, 40

Object space 5, 18, 83
Optical, *see also* Aberration *and*
 Aperture
 fill factor 140, 143
 invariant 9–11
 throughput 12
 transfer 9–12
 truncation; apodizing 52, 53
 ray directions 4, 5, 83, 161
 relay 11, 121, 137, 138
 schlieren method 156
 vignetting 141
Optics
 afocal 11, 140–142
 anamorphic 10, 112, 150–152
 binary; binary levels 143
 etching 143
 field lens 140–142
 fill factor 142, 143
 nodal center 7
 tandem 10
 telecentric 6, 121
Optical scanning
 analogy 2
 coherent, incoherent xii
 definition xi, 1, 2
Output system 8, 9
Oversampling 30, 31

Photodetector 6, 7, 20, 83
Point spread function (PSF)
 Gaussian illuminated aperture
  22–24
 intensity nulls 21, 22
 point object; delta function 27
 principal lobes 24, 25
 scanning; sliding variable
  24–28
 stationary function 25
 symmetry of function 25
 uniformly illuminated aperture
  20–22
Polarization 6, 80, 81, 96, 97
Polygon, *see* Scanning devices
Prism bow compensation 97

Prism rotators
 Dove 82
 Pechan 82
Programming
 electrical 138, 145
 positioning 145

Radiation, scattered 8, 147, 155, 158
Radial symmetry 15, 16, 18, 57–59,
 75–80, 85, 87, 101
 criteria 58, 77
 definition 58
 with tilted optical axis 77–79
Raster formation 27, 28, 101
 in television 27, 28
Rayleigh range 32–34
Reading, scanned 19
Relay, optical 11, 121, 137, 138
Reciprocal paths 3
Remote sensing, *see also* Scanning
 xii, 3, 8–11, 61, 62
Resolution distinctions
 astronomy 45
 hard copy 10
 photography 10, 46
 television 10, 46
Resolution, optical scanning, *see also*
 Apertue shape factor 9–12,
 45–62, 70, 73, 105, 107, 112,
 116
 acoustooptic 117
 augmented 48, 56–61
 equations 47, 48, 57, 59, 117
 holographic scanner 60
 invariant; determined at scanner
  9–12, 45, 48, 54, 73, 161
 nomogaph 48, 49, 162
 phased array 135
 prismatic polygon 70
 quadrature components 150
 remote sensing 61, 62
Resonant scanner, *see* Scanning
 devices, oscillatory
Retroreflection; retrocollection 4–7,
 94

Scanning, *see also* Scanning devices
    active 3–5, 8, 9, 15, 18, 161
    along-scan 28
    angular 16, 46, 48
    arced; curved 15–17, 87
    bandwidth 10, 70, 162
    bow 7, 8, 66, 97, 110, 111
    continuous lone 25
    cross-scan 28, 112
    descanning 6, 94
    double-pass 4, 7, 8, 74, 75, 163
    duty cycle 67, 68, 100, 101, 133
    error correction, *see* Beam
        misplacement correction
    external drum 17
    flying spot 3, 8, 18, 83, 117, 120
    format 12
    image rotation, derotation, *see*
        Image
    internal drum 16, 66, 87, 100, 107
    linearity 13, 24
    objective 4, 13, 16, 17
    passive 3, 4, 8, 9, 18, 83–85, 161,
        162
    postobjective 4, 5, 7, 15, 16, 108
    preobjective 4, 5, 13–15, 108
    quality criteria, *see also*
        Modulation transfer function
        depth of field; of focus 16, 36, 37
        spot size 11, 26, 27
    quantized; digital; sampling,
        see also Oversampling 27–31
    remote sensing 83–85, 162
    spatial, temporal xi, 1
    stitching 113
    theory 19–31, 161
        PSF movement 25–27
        sliding variable 25
    translational 4, 16, 46, 48
Scanning devices, *see also* Scanning
    Acoustooptic scanner 113–124,
        163
        acoustic absorber 114
        acoustic phased array 115
        acoustooptic effect 113

chirp system, *see* Chirp deflector
    diffraction efficiency 115
    materials, acoustooptic 113, 114
    modulation, *see also*
        Modulation 113, 117
    piezoelctric transducer 114, 121,
        214
Scophony system, *see* Scophony
    process
spectral coverage 113
transit time 116
traveling lens system, *see*
    Traveling lens
Agile beam steering 3, 105,
    128–145, 162
    analogies, refractive prism 130,
        131
    blazed grating, effective 139,
        140, 142, 144, 163
    electrooptic cell array 130–134,
        137
    decentered microlens arrays
        139–144, 162, 163
    decentering methods 140–143
    dielectric layer array 137
    diffraction efficiency 132, 133,
        137, 140
    field lens 140–142
    liquid crystal phase retarders
        133, 137
    microlens array 140–144
    mirrored piston array 130, 138,
        139
    motivation, principal 128
    origin; microwave-optical 129,
        132, 134
    output beam angle 134
    phased array resolution
        134–137
    phased array technology
        129–139, 162, 163
    PLZT material, *see also*
        Electrooptic 138
    spurious components 140,
        141

tandem or cascaded scanners 137, 138
techniques, principal 129
two-lens method 140, 141
Chirp deflector 123, 124, 163
  diffractive traveling lens 123
  filled aperture 124
  lens focal length 123
  pre-scanned aperture 124
Classification, family tree 63, 162
  inertia, high 63, 65–100, 103, 112, 113, 162
  inertia, low 63, 64, 101, 102, 112–145, 162, 163
  random access; agile 63, 113–145
  rotational, translational 64
  sawtooth ramp 64, 101
  sinusoidal (harmonic) scan 64, 102, 103, 106
Comparison of major types 164–167
  agile beam steering 144, 145, 164
  alternate techniques 167
  chart: acoustooptic, electrooptic, galvanometric, holographic, polygonal, resonant 165
Design considerations, polygon, *see also* Scanners, Polygon 69–85
  architectural factors 69–83
  double-pass operation 4, 7, 8, 74, 75, 163
  duty cycle 67, 68, 70, 74, 100
  facet tracking 65, 74, 75
  overfilled; underfilled 75
  postobjective operation 76–82
  preobjective operation 70–74
  prismatic polygon, design of 70–74
  pupil relief distance, *see also* Architecture 72, 73
  scan angle, Θ 70–73

scanner-lens relationships 72–74
Electrooptic scanners 63, 124–128
  broadband negative feedback 127
  convergent beam aberration 127
  convergent beam propagation 127, 163
  deflection development 124–126
  drive power dissipation 128
  electrooptic prisms 127
  gradient deflector 124
  material factors 126, 128
  quadrapolar electrodes 126
  ray propagation 124
  resolution equation 126
  time delay, low 127
  time dependent gradient 126
Fiberoptic scanner 106, 107
  ball lens 106, 107
  cantilevered fiber 106
  lissajous pattern 106
  resolution, scanned 107
Holographic scanner 85–100, 162, 165
  advantages 86, 87
  analogous arrangements 87
  analogous Bragg diffraction and prism refraction 90–92
  Bragg angle 86, 88, 96, 97
  characteristics 85, 87–90
  diffractor replication 86, 99
  disc wobble error 89, 90
  grating; linear, lenticular 85, 88, 89, 100
  grating equation 88
  Holofacet scanners 93
  Hologon scanners 97
  limitations 87
  multifunction scanners 97–100
  reflective scanners 92
  substrate options 93–98, 114
  substrate geometry 89

Oscillatory (vibrational) 64, 68,
    100–107, 145, 162, 163
  cross-coupling 104
  D'Arsonval movement 101
  damping 101–124
  duty cycle 103, 104
  fast steering mirror 100, 105,
    106, 112, 163
  flexure suspension 104, 105
  galvanometer 100, 101, 112, 113,
    145, 163
  linearization, resonant scanner
    103, 104
  mechanical Q 103
  mirror materials 105
  moving coil 101
  piezoelectric actuator 100, 106,
    114, 142
  pixel timing; modulation 103
  position sensing 104, 105
  radial stiffness 104
  random access 101
  resonant scanner 100–107, 162,
    163
  spurious modes 104
  suspension systems 101, 104,
    105
  torque 101, 102
  two-axis device 105
  voice coil actuator 105
Monogon 14, 15, 76, 80, 82, 100
Polygon, *see also* Scanning
    devices, Design
    considerations, polygon 162
  facet shift; facet "lift-up" 74, 79
  facet-to-axis angle 65, 66
  facet-to-facet angle 65, 66
  overfilling 52, 68, 69, 74
  prismatic 14, 58, 59, 65–76, 84,
    85
  pyramidal 14, 52, 58, 65–67,
    77–81, 108

  tilted lens axis 77–79
  underfilling 68, 69, 74, 81
Scophony process xii, 43, 117–119,
    163
  acoustic modulation 118, 119
  auxiliary deflector 118, 119
  coherence effects 119
  schlieren optics 119
  Scophony TV 117
  zero-order beam stop 118, 119
Traveling lens 119–123, 163
  acoustooptic lens formation
    119–121
  beam compressor 121
  cylindrical lensing 121
  harmonic pressure wave 121
  parabolic index change 121
  prescan deflector 120
  synthetic traveling lens 121, 123
Scan magnification 57–61, 76, 101
Scatter 6, 147, 155, 159
Scophony process, *see* Scanning
    devices
Spectral filtering 95
Specular component 6
Spot size, *see also* Point spread
    function 12, 24, 27, 43, 47, 50
Summary 161–164

Telescope 11, 142
  Galilean, Keplerian 142
TEM$_{00}$ mode 22
Transducers, drive
  piezoelectric 106, 114, 142, 199
  electrodynamic 101, 102, 105, 142
Traveling lens, *see* Scanning devices

Unification xii, 2, 4
  topics 2, 3
Uniform illumination, *see* Aperture

Writing, scanned 2, 19